A Course on Damage Mechanics

Springer-Verlag Berlin Heidelberg GmbH

Jean Lemaitre

A Course on Damage Mechanics

With a Foreword by
Professor H. Lippmann

Second Revised and Enlarged Edition

With 118 Figures

Springer

Professor Jean Lemaitre

Laboratoire de Mécanique et Technologie
E.N.S. de Cachan, C.N.R.S.
Université Paris 6
61, avenue du Président Wilson
F-94235 Cachan Cedex, France

ISBN 978-3-540-60980-3

Library of Congress Cataloging-in-Publication Data

Lemaitre, J. (Jean), 1934–
A course on damage mechanics / Jean Lemaitre ; with a foreword by H. Lippmann. – 2nd and enl. ed.
p. om.
Includes bibliographical references and index.
ISBN 978-3-540-60980-3 ISBN 978-3-642-18255-6 (eBook)
DOI 10.1007/978-3-642-18255-6
1. Fracture mechanics, 2. Continuum damage mechanics, I. Title.
TA409.L46 1996
620.1'126–dc20

Originally published by Springer-Verlag Berlin Heidelberg in 1996

Typesetting: Fotosatz-Service Köhler OHG, Würzburg
Cover design: de'blik, Berlin
SPIN: 10710372 55/3012 - 5 4 3 2 1 - Printed on acid-free paper

Foreword

A new branch of science usually develops thus. Somebody publishes the basic ideas. Hesitatingly at first, then little by little, other original contributions appear, until a certain threshold is reached. Then, overview articles are printed, conferences are held, and a first mention is made in textbooks, until specialized monographs are written. Continuum damage mechanics has reached that status now.

To analyze or, if possible, to predict the failure of machine parts or other structures is one of the main goals of engineering science. Consequently fracture mechanics became one of its leading branches. It was based on the analysis of existing cracks. However, especially under conditions of cyclic loading, this might be too late to prevent a disaster. Therefore, the question regarding the precursory state, that is, the evolution of internal damage before macrocracks become visible, was then posed. One of the successful approaches to the problem was Weibull's theory which examined, in a statistical manner, the "weakest link" in the material volume under consideration. Unfortunately it proved too difficult mathematically to be applied to complicated parts or structures. Therefore it was highly appreciated by the scientific community when L.M. Kachanov published in 1958 a simple model of material damage which subsequently could be extended to brittle elastic, plastic or viscous materials under all conditions of uniaxial or multiaxial, simple or cyclic loadings, so that it may be considered nearly universal. As usual in a living science many questions are still open, and the limitations of the model need further research.

Jean Lemaitre, the author of the present work, has become one of the most distinguished representatives of damage mechanics. His monograph highlights all the aspects of the subject, from the basic ideas, to theoretical and thermodynamical perspectives, up to numerical or experimental applications. It also devotes sufficient space to the problems associated with parameter identification. I like the style of presentation, which is always vivid or humorous and never dry, so that the understanding even of the more difficult passages is facilitated. An excellent book has finally appeared, most valuable to any engineer or scientist dealing with modern problems of strength of materials.

I wish it every success.

Munich, 6 August 1990.

Professor Horst Lippmann
Lehrstuhl A für Mechanik
Technische Universität München,
Germany

Introduction

Why an apple on the cover page?
And why not? Do you know any other fruit species with so many different flavors? Besides, I like apples! It is also a joke from the early 1970s when it was so fashionable in France to teach "modern" mathematics. At that time you could draw a potato on the blackboard as the set of points having the neighbouring Bernoulli's property of representing a beam! So, to tease my colleagues, I began to draw an apple as an elementary solid, and the students just loved it!
Why damage?
Since people began to design and make components most of their efforts have been spent trying to avoid failure! Stress and strain concepts were hardly developed before the concept of damage appeared; this was probably because they are related to displacement, which can be seen and touched. Damage is the deterioration which occurs in materials prior to failure, and generally you see and you touch nothing! For centuries fracture has been studied as a "yes or no" process related to a critical value of load, stress, strain, and time or number of cycles of loading. A. Palmgreen (1924), M.A. Miner (1945), and E.L. Robinson (1952) pioneered the concept of a variable related to the progressive deterioration prior to failure, but 1958 is the year to be considered the starting point of continuum damage mechanics, when L.M. Kachanov published the first paper on a field variable ψ called "continuity". About fifteen years later $D = (1 - \psi)$ received the status of an internal state variable in the thermodynamical sense: $0 \leq D \leq 1$ (0 for the undamaged state and 1 for failure). During these fifteen years, this concept was practically ignored and only one important result appeared, in 1968, with the concept of effective stress introduced by Y.N. Rabotnov. The basic developments of damage mechanics occurred during the 1970s, at least ten years after the tremendous development of fracture mechanics. In the 1980s, the theory was set up on a more rigorous basis using thermodynamics and micromechanics, and applications to engineering began as many more people were involved in this discipline.
Why a course?
The idea of assembling, into a coherent instructional unit, the main results of damage mechanics came from Prof. Lee Hao when he invited me to teach a course in 1983 at Huazhong University in Wuhan (China). Afterwards, it became the "Damage Circus" which travelled to Rio de Janeiro (Brazil) in 1985 invited by Prof. R. Sampaio, to Nagoya (Japan) in 1986 invited by Prof. S. Murakami, to München (Germany) in 1987 invited by Prof. H. Lippmann and based in Paris for a graduate course "Mécanique de l'endommagement" of a joint "Université-

Grandes Ecoles'' 3rd cycle ''Mécanique et matériaux'' initiated by D. François, A. Pineau, and A. Zaoui. The opportunity to write ''Damage Mechanics'' in book form was offered to me by the Humboldt Foundation through an award which gave me the possibility to spend some ''free time'' in 1988 and 1989 with Prof. H. Lippmann and Prof. J. Najar at the Technische Universität Munchen. The second opportunity to put pen to paper was provided by an invitation of Prof. F. Leckie to spend the whole academic year, 1989–90 at the University of California, Santa Barbara (USA) where I benefited from the stimulating company of F. Leckie, A. Evans, J. Hutchinson and R. McMeeking. By the way, the ''Damage Circus'' also played as the ''Damage Brothers'' in 1986 for a CISM course in Udine (Italy) together with my friends J. Hult, D. Krajcinovic, F. Leckie, and S. Murakami and as the ''Modern Damage Quartet'' in 1988 and 1990 with my young colleagues and friends A. Benallal, R. Billardon and D. Marquis. For this synthesis, which is certainly neither complete nor accurate, I have benefited from several conferences devoted to damage mechanics listed in the literature references. The French group of labs ''GRECO-CNRS'' ''Grandes deformations et endommagement'', which included more than 25 groups working together between 1980 and 1989, was the melting pot of the French solid mechanics school initiated by P. Germain, in which many ideas and applications were discussed. But I must say that it was first in ONERA with my friend J.L. Chaboche and later, after 1975, at the ''Laboratoire de Mécanique et Technologie'' at Cachan, that I found an exceptional atmosphere of research and friendship in which to discuss, develop or reject ideas. I thank you all for your help and your friendship.

Why four chapters?

The main purpose of this book is to develop mechanics, which is a macroscopic modelling, so that it can be used in the design, processing, testing and control of mechanical components in engineering. Then it becomes physics for mechanicans, mechanics for mathematicians, mathematics for computer specialists, computations for engineers, and engineering for physicists!

- The first chapter deals with the phenomenological aspects of damage with a description of the different kinds of damage and the main ideas of modelling applied to the one-dimensional case. An important section is devoted to damage measurements, the field which probably needs most further research.
- The second chapter concerns the main part of damage mechanics in which the couplings between damage and strains are written for the three-dimensional case. Thermodynamics is the main framework of the theory, although micromechanics is widely used to derive the analytical form of the general functions introduced.
- The whole third chapter is devoted to the kinetic laws of damage evolution. It is an attempt to unify many models that have been proposed for brittle damage, ductile and creep damage, low and high cycle fatigue of metals, polymers, concrete, ceramics and composites.
- In the fourth chapter, several methods are described for predicting the conditions for crack initiation in structural components: either simple methods with a few calculations or more accurate methods involving computerized calculations, and

the finite element method. A post-processor of damage growth, up to crack initiation, is described in detail.

These chapters are written as a course for the senior-undergraduate or graduate level with detailed calculations and exercises at the end of each chapter section. This is a course for universities or ''Grandes Ecoles'' but also, I hope, a new course for the permanent education of engineers in research centers and industry.

Many people have read the manuscript to improve shortcomings in my English and my explanations, so let me then thank them for their advice: Prof. H. Lippmann who wrote the Foreword, Prof. F. Leckie, Drs. A. Benallal, R. Billardon, I. Doghri and D. Marquis; some of my students at UC-Santa Barbara: S. Gunawardena, K. Herath, J. Manderscheid, D. Sherman; and June Finney and Sabine Lemaitre who typed the manuscript with much care. The cover picture was taken by Annie Lemaitre. Merci.

Spring 1988 – Spring 1991. Jean Lemaitre

Contents

Notation xv

Chapter 1 Phenomenological Aspects of Damage 1

1.1 Physical Nature of the Solid State and Damage 1
1.1.1 Atoms, Elasticity and Damage 1
1.1.2 Slips, Plasticity and Irreversible Strains 3
1.1.3 Scale of the Phenomena of Strain and Damage 3
1.1.4 Different Manifestations of Damage 4
1.1.5 Exercise on Micrographic Observations 10

1.2 Mechanical Representation of Damage 11
1.2.1 One-Dimensional Surface Damage Variable 11
1.2.2 Effective Stress Concept 12
1.2.3 Strain Equivalence Principle 13
1.2.4 Coupling Between Strains and Damage; Rupture Criterion; Damage Threshold 14
1.2.5 Exercise on the Micromechanics of the Effective Damage Area 17

1.3 Measurement of Damage 19
1.3.1 Direct Measurements 19
1.3.2 Variation of the Elasticity Modulus 21
1.3.3 Variation of the Microhardness 25
1.3.4 Other Methods 28
1.3.5 Exercise on Measurement of Damage by the Stress Amplitude Drop 36

Chapter 2 Thermodynamics and Micromechanics of Damage 39

2.1 Three-Dimensional Analysis of Isotropic Damage 39
2.1.1 Thermodynamical Variables, State Potential 39
2.1.2 Damage Equivalent Stress Criterion 44
2.1.3 Potential of Dissipation 46
2.1.4 Strain-Damage Coupled Constitutive Equations 50
2.1.5 Exercise on the Identification of Material Parameters . . 59

2.2 Analysis of Anisotropic Damage 59
2.2.1 Geometrical Definition of a Second-Order Damage Tensor 60
2.2.2 Thermodynamical Definition of a Fourth-Order Damage Tensor 63
2.2.3 Energetic Definition of a Double Scalar Variable 64
2.2.4 Exercise on Anisotropic Damage in Proportional Loading 66

2.3 Micromechanics of Damage 68
2.3.1 Brittle Isotropic Damage 68
2.3.2 Ductile Isotropic Damage 72
2.3.3 Anisotropic Damage 75
2.3.4 Microcrack Closure Effect, Unilateral Conditions 80
2.3.5 Damage Localization and Instability 86
2.3.6 Exercise on the Fiber Bundle System 90

Chapter 3 Kinetic Laws of Damage Evolution 95

3.1 Unified Formulation of Damage Laws 95
3.1.1 General Properties and Formulation 96
3.1.2 Stored Energy Damage Threshold 98
3.1.3 Three-Dimensional Rupture Criterion 100
3.1.4 Case of Elastic-Perfectly Plastic and Damageable Materials 105
3.1.5 Identification of the Material Parameters 109
3.1.6 Exercise on Identification by a Low Cycle Test 110

3.2 Brittle Damage of Metals, Ceramics, Composites and Concrete 112
3.2.1 Pure Brittle Damage 113
3.2.2 Quasi-brittle Damage 113
3.2.3 Exercise on the Influence of the Triaxiality on Rupture 116

3.3 Ductile and Creep Damage of Metals and Polymers 117
3.3.1 Ductile Damage 117
3.3.2 Exercises on the Fracture Limits in Metal Forming . . . 119
3.3.3 Creep Damage 123
3.3.4 Exercise on Isochronous Creep Damage Curves 126

3.4 Fatigue Damage 127
3.4.1 Low Cycle Fatigue 127
3.4.2 Exercise on Creep Fatigue Interaction 132
3.4.3 High Cycle Fatigue 135
3.4.4 Exercise on Damage Accumulation 141

3.5 Damage of Interfaces 143
3.5.1 Continuity of the Stress and Strain Vectors 143
3.5.2 Strain Surface Energy Release Rate 145

3.5.3 Kinetic Law of Debonding Damage Evolution 146
3.5.4 Simplified Model 147
3.5.5 Exercise on a Debonding Criterion for Interfaces 150

3.6 Table of Material Parameters 152

Chapter 4 Analysis of Crack Initiation in Structures 154

4.1 Stress-Strain Analysis 154
4.1.1 Stress Concentrations 156
4.1.2 Neuber's Method 163
4.1.3 Finite Element Method 165
4.1.4 Exercise on the Stress Concentration Near a Hole . . . 167

4.2 Uncoupled Analysis of Crack Initiation 169
4.2.1 Determination of the Critical Point(s) 169
4.2.2 Integration of the Kinetic Damage Law 169
4.2.3 Exercise on Fatigue Crack Initiation Near a Hole 172

4.3 Locally Coupled Analysis 176
4.3.1 Localization of Damage 176
4.3.2 Postprocessing of Damage Growth 177
4.3.3 Description and Listing of the Postprocessor DAMAGE 90 179
4.3.4 Exercises Using the DAMAGE 90 Postprocessor 196

4.4 Fully Coupled Analysis 201
4.4.1 Initial Strain Hardening and Damage 202
4.4.2 Example of a Calculation Using the Finite Element Method 203
4.4.3 Growth of Damaged Zones and Macrocracks 206
4.4.4 Exercise on Damage Zone at a Crack Tip 208

4.5 Statistical Analysis with Microdefects 209
4.5.1 Initial Defects 210
4.5.2 Case of Brittle Materials 210
4.5.3 Case of Quasi Brittle Materials 212
4.5.4 Case of Ductile Materials 214
4.5.5 Volume Effect 215
4.5.6 Effect of Stress Heterogeneity 217
4.5.7 Exercise on Bending Fatigue of a Beam 219

Bibliography 222

History of International Damage Mechanics Conferences 223

Authors and Subject Index 225

Notation

The choice of notation is always a difficult problem in a new field because all letters have already been assigned a meaning. I have tried to be as classical as possible, but I could not completely avoid using the same letter for different meanings. Be careful and please forgive me.

Operators

x	a scalar
x_i	components of a vector $\vec{x}$
x_{ij}	components of a second-order tensor $\mathbf{x}$
$[x]$	matrix
$\dot{x}$	time derivative of $x (\dot{x} = dx/dt)$
$x_{i,j}$	gradient of $\vec{x}$
$x_{ij,j}$	divergence of $\mathbf{x}$
δ_{ij}	Kronecker delta: $\delta_{ij} = 1$ if $i = j$, $\delta_{ij} = 0$ if $i \neq j$
x_{kk}	trace of $\mathbf{x}$
$x_{ij}x_{ij}$	trace of $\mathbf{x}^2$
x_{ij}^D	components of the deviatoric tensor $x_{ij}^D = x_{ij} - \frac{1}{3}x_{kk}\delta_{ij}$
$\|x\|$	absolute value of x
$\mathrm{sign}(x)$	+ or − as the sign of x
$[\![x]\!]$	discontinuity of x
$\langle x \rangle$	Macauley bracket $\langle x \rangle = 0$ if $x < 0$, $\langle x \rangle = x$ if $x \geq 0$
$\overline{x}$	mean value of x
Δx	range of x (peak to peak amplitude)
d, ∂, δ	differential operators
H	Heaviside function: $H(x) = 0$ if $x < 0$, $H(x) = 1$ if $x \geq 0$
Ln	Napierian logarithm

Symbols

a	crack length
a_{ijkl}	elastic moduli
A	Gurson-Tvergaard material parameter
A	crack area
b	isotropic hardening material exponent
B	Gurson-Tvergaard material parameter

C	Paris' crack growth material parameter
d	length of microcells
D	damage variable
D_s, D_n	damage variables in shear and tension
$D_{ij}, \mathbf{D}$	components and second-order anisotropic damage tensor
D_{ijkl}	components of anisotropic fourth-order damage tensor
D_c	critical damage at crack initiation
D_{1c}	critical damage in pure tension
$\dot{e}$	specific internal energy rate
E	Young's modulus of elasticity
E_T	elastoplastic tangent modulus
$\tilde{E}$	effective damage modulus of elasticity, $\tilde{E}^+$ in tension, $\tilde{E}^-$ in compression
f	yield function
$\vec{f}$	body force
F	force
F	function potential of dissipation
F_D	damage potential function
g	strain-stress function
$\vec{g}$	vector of strain rate discontinuity
G	strain energy release rate
G_c	energy toughness of materials
h	crack closure parameter
H	microhardness
H	Heaviside function
H_{ijkl}	tangent elasto-plastic stiffness matrix components
i	intensity of electricity
I_{kl}	unit second-order tensor components
I_{ijkl}	unit fourth-order tensor components
k	factor
$\dot{k}$	specific kinetic energy rate
K	stress intensity factor
K_p	cyclic plasticity material parameter
K_v	creep material parameter
K_∞	viscosity material parameter
K_T	stress concentration coefficient

l	length of the mesovolume element
m	mass
M	material point
M	strain hardening material exponent
n	number of defects
n	viscosity material exponent
$\vec{n}$	unit normal vector
N	number of cycles
N	Norton's creep material exponent
N_R	number of cycles to rupture
p	accumulated plastic strain
p_D	damage threshold plastic strain
p_x	specific external force power
P	porosity
$\vec{q}$	heat flux vector
Q	input heat rate
r	resistivity
r	strain of isotropic hardening
R	isotropic hardening stress variable
R_∞	isotropic hardening material parameter
Ro, Ra	stiffness
$R\nu$	triaxiality function
s	specific entropy
s	crack area
S	sectional area
S_D	damaged sectional area
S	damage strength material parameter
t	time
T	temperature
$\vec{u}$	displacement vector
v	wave speed
V	electrical potential difference
V	volume
w_e	elastic strain energy density
w_e^*	elastic complementary energy density

W_e	elastic strain energy of a body
x	coordinate
X, X_{ij}^D	unidimensional and components of tensorial kinematic hardening stress variable
X_∞	kinematic hardening material parameter
Y	strain energy density release rate
Y_s, Y_n	strain energy density release rate in shear and tension
Y_{ijkl}	strain energy release rate tensor components
Y_c	critical strain energy density release rate at rupture
α, α_{ij}	unidimensional and components of tensorial variable of kinematic hardening
γ	nonlinear kinematic hardening material parameter
Δ	amplitude
$\varepsilon, \varepsilon_{ij}, \boldsymbol{\varepsilon}$	uniaxial and tensorial total strains
$\varepsilon_e, \varepsilon_{ij}^e, \boldsymbol{\varepsilon}^e$	uniaxial and tensorial elastic strains
$\varepsilon_p, \varepsilon_{ij}^p, \boldsymbol{\varepsilon}^p$	uniaxial and tensorial plastic strains
ε_H^e	hydrostatic elastic strain
ε_{eq}	equivalent total strain
ε_{p_D}	damage plastic strain threshold in pure tension
ε_R	strain to rupture
η	Paris' crack growth material exponent
ϕ	dissipation
λ	Lamé's coefficient of elasticity
$\dot{\lambda}$	plastic multiplier
μ	Lamé's coefficient of elasticity in shear
ν	Poisson's ratio of elastic contraction
ν^*	elastoplastic contraction coefficient
$\omega, \omega_{ij}, \omega_{ijkl}$	scalar and tensorial volume damage variables
ρ	mass density
$\sigma, \sigma_{ij}, \boldsymbol{\sigma}$	uniaxial and tensorial stresses

$\sigma^{\mu}, \sigma_{ij}^{\mu}, \boldsymbol{\sigma}^{\mu}$	stresses at microscale
σ_H	hydrostatic stress
$\sigma_{ij}^D, \boldsymbol{\sigma}^D$	deviatoric stress tensor
σ_{eq}	von Mises equivalent stress
σ^*	damage equivalent stress
$\tilde{\sigma}, \tilde{\sigma}_{ij}, \tilde{\boldsymbol{\sigma}}$	unidimensional and tensorial effective stresses
σ_f	fatigue limit stress
σ_y	yield stress
σ_s	plastic threshold stress
σ_u	ultimate stress
σ_R	stress to rupture
ψ	Helmholtz specific free energy

Chapter 1

Phenomenological Aspects of Damage

The damage of materials is the progressive physical process by which they break. The mechanics of damage is the study, through mechanical variables, of the mechanisms involved in this deterioration when the materials are subjected to loading. At the microscale level this is the accumulation of microstresses in the neighborhood of defects or interfaces and the breaking of bonds, which both damage the material. At the mesoscale level of the representative volume element this is the growth and the coalescence of microcracks or microvoids which together initiate one crack. At the macroscale level this is the growth of that crack. The two first stages may be studied by means of damage variables of the mechanics of continuous media defined at the mesoscale level. The third stage is usually studied using fracture mechanics with variables defined at the macroscale level.

When studying engineering materials such as metals and alloys, polymers and composites, ceramics, rocks, concrete, and wood, it is very surprising to see how such materials, which have different physical structures, are similar in their qualitative mechanical behavior. All show elastic behavior, yielding, some form of plastic or irreversible strain, anisotropy induced by strain, cyclic hysteresis loops, damage by monotonic loading or by fatigue, and crack growth under static or dynamic loads. This means that the common mesoscopic properties can be explained by a few energy mechanisms that are similar for all these materials. This is the main reason it is possible to explain material behavior successfully with the mechanics of continuous media and the thermodynamics of irreversible processes, which model the materials without detailed reference to the complexity of their physical microstructures.

1.1 Physical Nature of the Solid State and Damage

1.1.1 Atoms, Elasticity and Damage

All materials are composed of atoms, which are held together by bonds resulting from the interaction of electromagnetic fields. Elasticity is directly related to the relative movement of atoms. The physical study of the properties of an atomic lattice leads to the theory of elasticity, but a much easier way is to write the mathematical constitutive equations directly at the mesoscale level using the property of reversibility of strain, which implies a one-to-one relationship, and eventually incorporate the properties of linearity and isotropy.

When debonding occurs, this is the beginning of the damage process. For example, metals are organized in crystals or grains: a regular array of atoms except on many lines of dislocations where atoms are missing. If a shear stress is applied, the dislocations may move by the displacement of bonds, thus creating a plastic strain by slip without any debonding as shown in Figure 1.1.

If the dislocation is stopped by a microdefect or a microstress concentration, it creates a constrained zone in which another dislocation may be stopped. This second process cannot occur without a debonding damage as shown in Figure 1.2. Several arrests of dislocations nucleate a microcrack. Other damage mechanisms in metals are intergranular debonding and decohesion between inclusions and the matrix.

All these mechanisms create plastic microstrains.

- — In polymers, damage occurs by the breakage of bonds that exist between the long chains of molecules.
- — In composites damage is the debonding between the fibers and the polymeric matrix.
- — In ceramics it is mainly microdecohesions between the inclusions and the matrix.
- — In concrete, the early damage mechanism is also a decohesion between aggregates and the cement with the complex influence of water.
- — In wood, the weak point where damage occurs is the bonding of the celulosic cells.

In all cases elasticity is directly influenced by the damage, since the number of atomic bonds responsible for elasticity decreases with damage. This coupling,

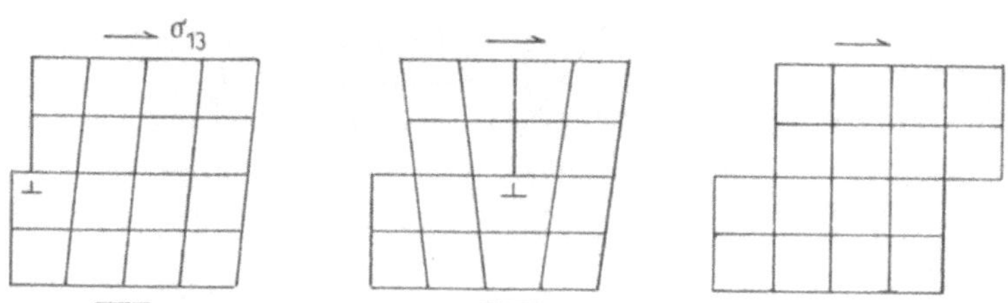

Fig. 1.1. Elementary plastic strain by slip due to dislocation movement

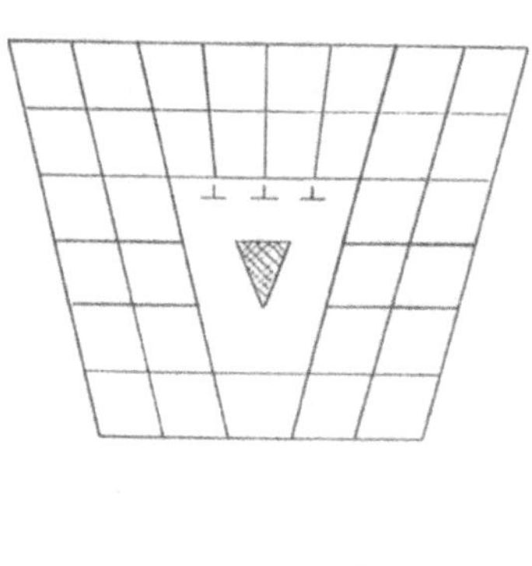

Fig. 1.2. Elementary damage by nucleation of a microcrack due to an accumulation of dislocations (after D. Krajcinovic)

which occurs at the level of the state of the material defined here by the elastic strain and the damage, is called a ''state coupling''.

1.1.2 Slips, Plasticity and Irreversible Strains

Plasticity is directly related to slips. In metals, slips occur by movement of dislocations, as shown schematically in Figure 1.1 or by the climbing of dislocations and twinning. In no case does it induce any appreciable volume change.
In other materials, irreversible strains may occur by different mechanisms:

— rearrangement of molecules in polymers;
— microcracks in ceramics where the large lattice resistance does not allow movements of dislocations;
— slips along surfaces of decohesions in concrete;
— rearrangement of cells in wood.

They may induce a volume change.
In all cases, damage influences plastic or irreversible strains only because the elementary area of resistance decreases as the number of bonds decreases. The damage does not directly influence the mechanism of slip itself; that is, there is no state coupling. The indirect coupling owing to an increase in the effective stress arises only in the kinetic constitutive equation, it is called ''kinetic coupling''.

1.1.3 Scale of the Phenomena of Strain and Damage

— Elasticity takes place at the level of atoms.
— Plasticity is governed by slips at the level of crystals or molecules.
— Damage is debonding from the level of atoms to the mesoscale level for crack initiation.

Continuum mechanics deals with quantities defined at a mathematical point. From the physical point of view, these quantities represent averages on a certain volume. The ''Representative Volume Element'', must be small enough to avoid smoothing of high gradients but large enough to represent an average of the microprocesses. For experimental purposes and numerical analysis it is useful to consider the following orders of magnitude of the Representative Volume Element which is the scale of mesomechanics:

– metals and ceramics:	$(0.1\ \text{mm})^3$
– polymers and most composites:	$(1\ \text{mm})^3$
– wood:	$(10\ \text{mm})^3$
– concrete:	$(100\ \text{mm})^3$

Another important property to consider is that the damage is always much more localized than the strain. Remember that the damage, or debonding of atoms, is

restricted to a surface, although the strains, being movements of atoms by variation of their distance or by many slips, occur throughout the volume. If damage exists in a single plane at the mesoscale, there is no way to study it by classical continuum mechanics. Fortunately, most often it exists on many planes at the microscale, but always with a high space gradient. Micromechanics helps a great deal in for modeling the damage behavior, as we shall see in Chapter 2.
In conclusion:

— the microscale is the scale of the mechanisms used to consider strains and damage;
— the mesoscale is the scale at which the constitutive equations for mechanics analysis are written;
— the macroscale is the scale of engineering structures.

1.1.4 Different Manifestations of Damage

Even if the damage at the microscale is governed by one general mechanism of debonding, at the mesoscale it can manifest itself in various ways depending upon the nature of the materials, the type of loading, and the temperature.

Brittle damage

The damage is called brittle when a crack is initiated at the mesoscale without a large amount of plastic strain. Just to give an order of magnitude, let us say that the ratio of plastic strain to elastic strain is below unity:

$$\frac{\varepsilon_p}{\varepsilon_e} < 1$$

This means that the cleavage forces are below the forces that could produce slips but are higher than the debonding forces. The degree of localization is high. An example is given in Figure 1.3.

Ductile damage

On the other hand, the damage is called ductile when it occurs simultaneously with plastic deformations larger than a certain threshold p_D. It results from the nucleation of cavities due to decohesions between inclusions and the matrix followed by their growth and their coalescence through the phenomenon of plastic instability. As a consequence, the degree of localization of ductile damage is comparable to that of plastic strain. An example is given in Figure 1.4.

Creep damage

When a metal is loaded at elevated temperature, for instance a temperature above 1/3 of the melting temperature, the plastic strain involves viscosity; that is, the material may be deformed at constant stress. When the strain is large enough, there

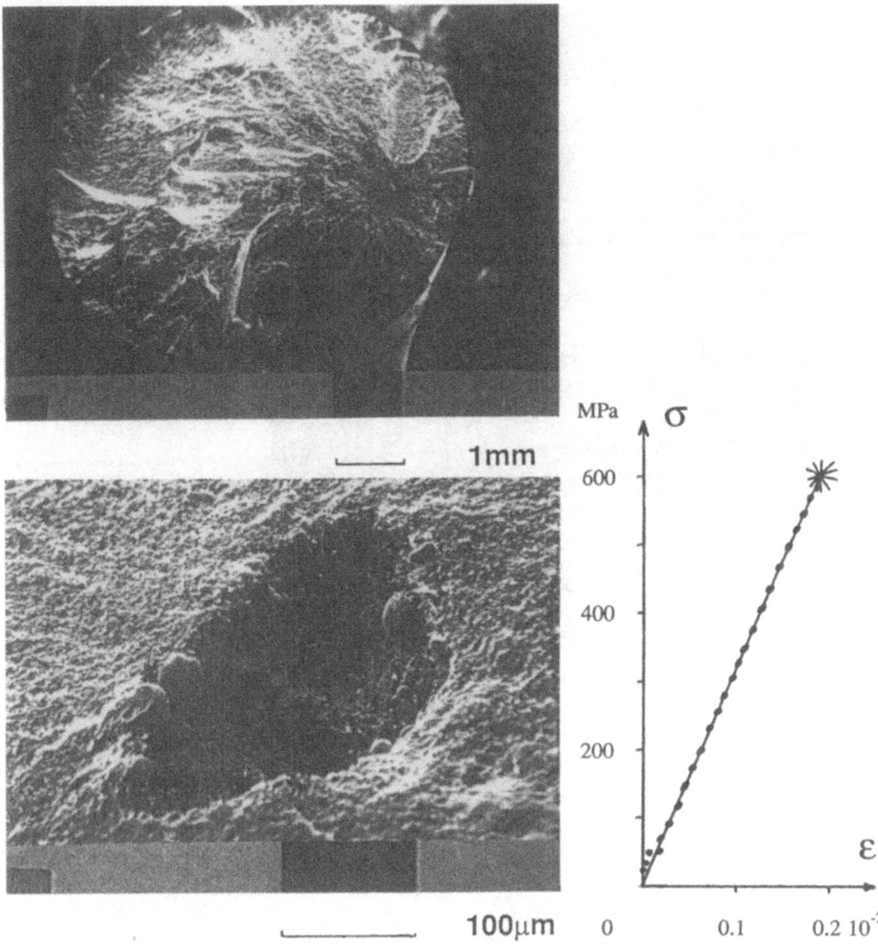

Fig. 1.3. a Fractograph showing microcracks damage around an inclusion in a silicon nitride ceramic; **b** Tensile stress-strain curve up to rupture (after F. Hild)

are intergranular decohesions which produce damage and an increase of the strain rate through the period of tertiary creep. As for ductile damage, the gradients of creep damage are similar to the visco-plastic strain gradients. An example is given in Figure 1.5.

Low cycle fatigue damage

When a material is subjected to a cyclic loading at high values for stress or strain, damage develops together with cyclic plastic strain after a period of incubation preceding the phases of nucleation and propagation of microcracks. The degree of

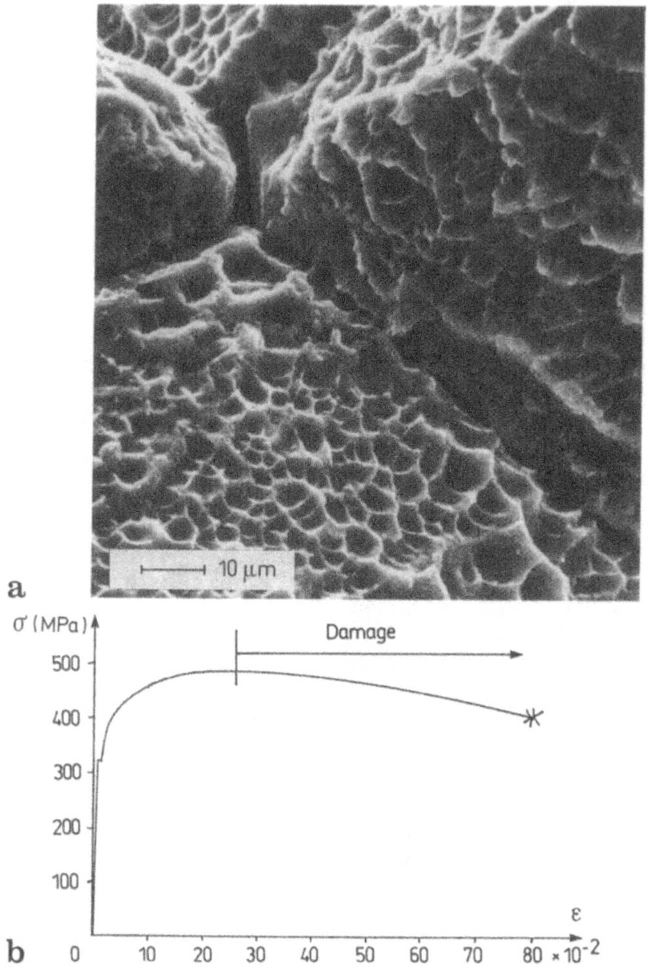

Fig. 1.4. **a** Fractograph showing microcavities damage at the grain boundaries of a steel (after L. Engel and H. Klingele); **b** Tension stress-strain curve of AISI 1010 annealed steel (after J. Dufailly)

damage localization is higher than for ductile or creep damage. Because of the high values for the stress, the low cycle fatigue is characterized by low values of the number of cycles to rupture, N_R:

$$N_R < 10000 \text{ cycles.}$$

If the material is strain loaded, the damage induces a drop of the stress amplitude as shown in Figure 1.6 for two stress-strain loops corresponding to the stabilized cycle and a cycle close to the rupture.

a b

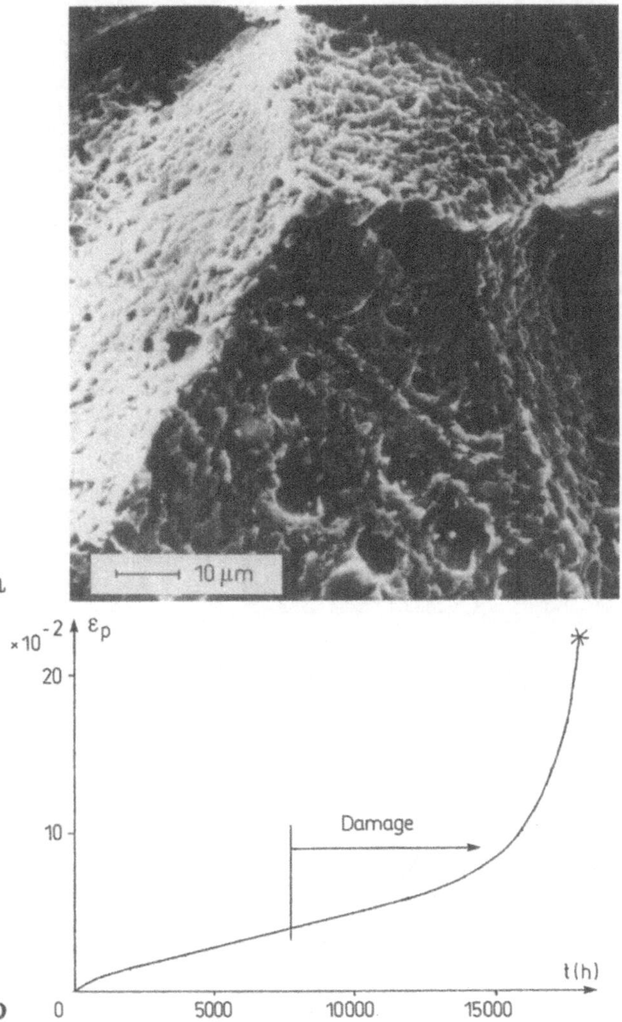

Fig. 1.5. **a** Fractograph showing intergranular cavitation damage formed at 800°C in a nickel-based alloy (after L. Engel and H. Klingele); **b** Creep curve under constant tension stress of A 542 stainless steel at 550°C (after A. Benallal)

For metals, the damage can be either intergranular or transgranular microcracking following slip-band arrests.

High cycle fatigue

When a material is loaded with lower values for stress, the plastic strain at the mesolevel remains small and is often negligible. It may be high at some points at

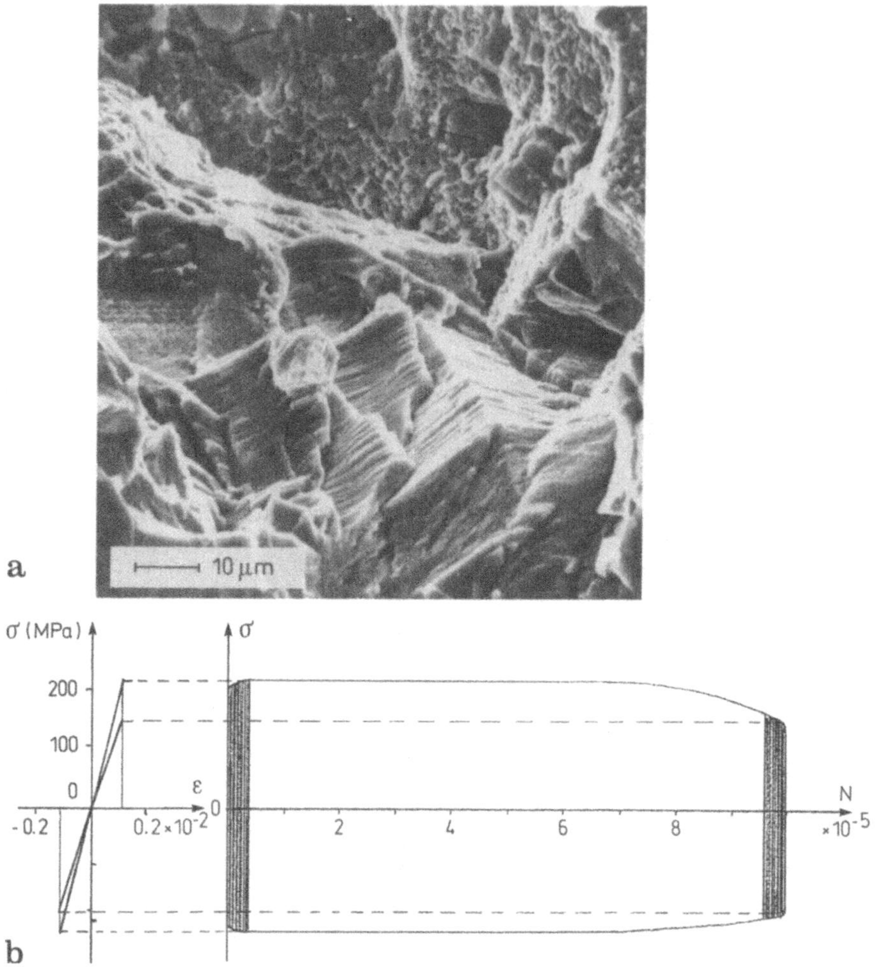

Fig. 1.7. a Fractograph of transcrystalline fatigue fracture in a nickel alloy (after L. Engel and H. Klingele); **b** Cyclic tension compression curves for high cycle fatigue of A 316 stainless steel

Note that for brittle damage and high cycle fatigue damage, a stress-strain curve obtained from a classical tension-compression test at the meso-scale usually does not represent the "true" behavior for strain and damage because the space localization induces microplastic and damage zones much smaller than those of the specimens. Nevertheless, it is used because mechanical experiments at the microscale are difficult to perform; but please bear in mind that the results are averages of nonuniform quantities over a mesovolume. The microhardness test may help to characterize a microvolume as it involves a size of the order of microns but its state of stress is complex.

a

b

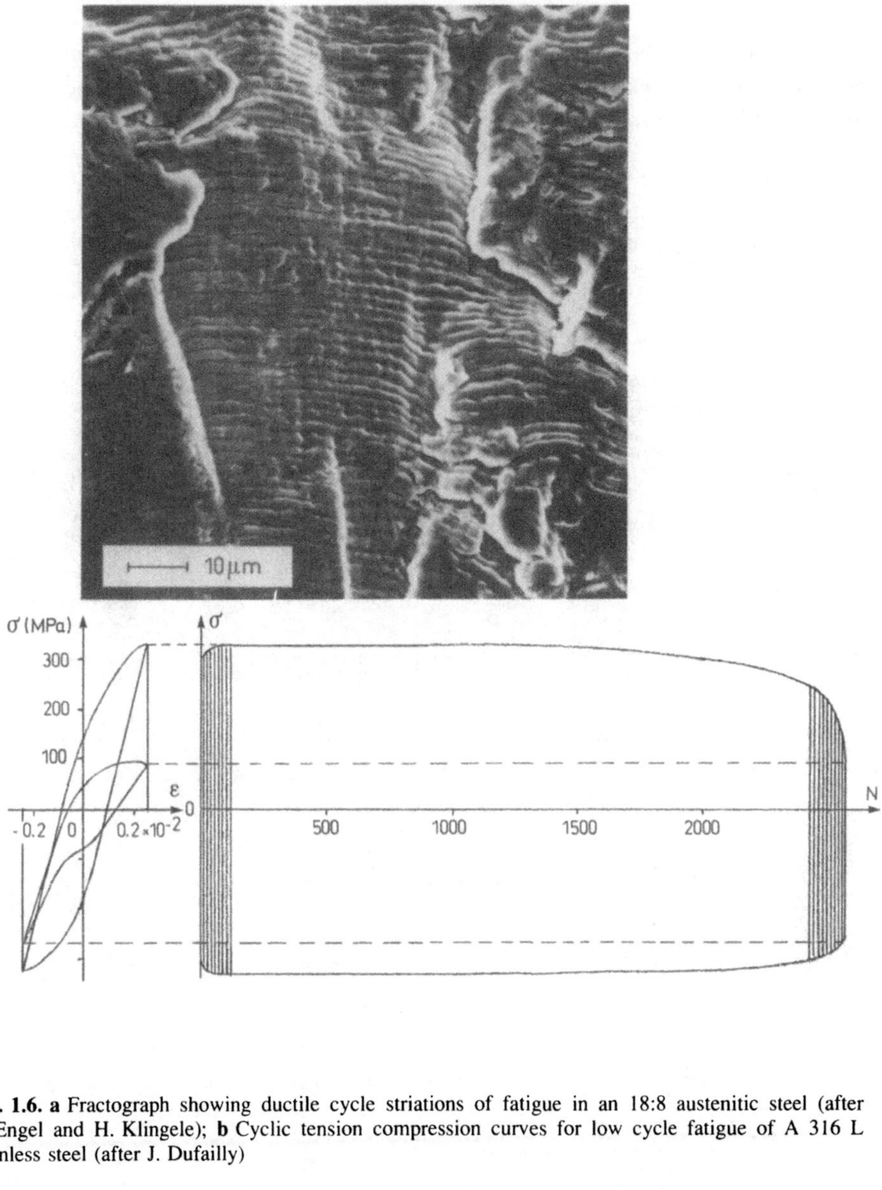

Fig. 1.6. a Fractograph showing ductile cycle striations of fatigue in an 18:8 austenitic steel (after L. Engel and H. Klingele); **b** Cyclic tension compression curves for low cycle fatigue of A 316 L stainless steel (after J. Dufailly)

the microlevel where transgranular microcracking occurs only on some planes and most often at the surface of the specimen by the mechanism of intrusion-extrusion. The number of cycles to failure may be very large:

$$N_R > 100000 \text{ cycles.}$$

As a consequence, the localization of damage is high and the drop of stress at the mesolevel for an experiment similar to that of Figure 1.6 occurs later on Figure 1.7.

1.1.5 Exercise on Micrographic Observations

Look carefully at the pictures of materials in Figure 1.8 taken with a microscope and *find the mechanism of strain damage or rupture for each case*. The answers are, in random order:

A— Transcristalline fatigue crack propagation in a nickel alloy at 650°C
C— Brittle transcristalline cleavage fracture in pure tantalum at room temperature
A— Creep voids at the grain boundaries of a nickel-based alloy at 800°

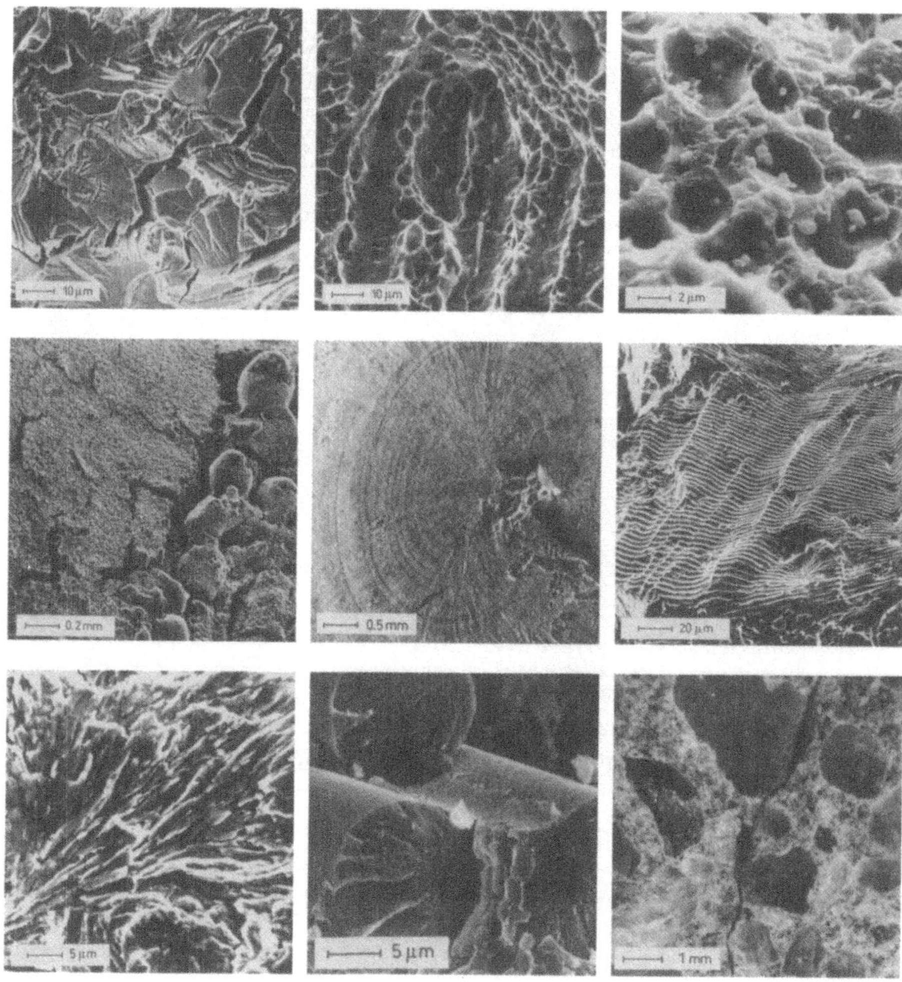

Fig. 1.8. Micrograph pictures taken from An Atlas of Metal Damage by L. Engel and H. Klingele and also from D. François and Y.Berthaud for the two last pictures

N – Fracture of a glass fiber epoxy composite
P – Fatigue crack in a steel specimen at room temperature
E – Tensile crack in a concrete
H – Ductile dimple fracture around manganese sulphide stringers in a steel at room temperature
M – Thermal cycling cracks in the nozzle of a metal squirting device
G – Stress corrosion transcristalline cracking in an austenitic stainless steel.

Hint: if the answers are correctly matched to the pictures, the sequence of capital reference letters should spell the name of a most famous wine.

1.2 Mechanical Representation of Damage

1.2.1 One-Dimensional Surface Damage Variable (L.M. Kachanov, 1958)

It follows from the preceding section that damage may be interpreted at the microscale as the creation of microsurfaces of discontinuities: breaking of atomic bonds and plastic enlargement of microcavities. At the mesoscale, the number of broken bonds or the pattern of microcavities may be approximated in any plane by the area of the intersections of all the flaws with that plane. In order to manipulate a dimensionless quantity, this area is scaled by the size of the representative volume element. This size is of primary importance in the definition of a variable continuous in the sense of continuum mechanics. At one point, it must be the representative effect on failure of microdefects over the mesoscale volume element. It is similar to plasticity where the plastic strain ε_p represents, at one point, the average of many slips.
Let us consider a damaged body and a Representative Volume Element (RVE) at a point M oriented by a plane defined by its normal $\vec{n}$ and its abscissa x along the direction $\vec{n}$ Figure 1.9.

— Let δS be the area of the intersection of the plane with the RVE;
— Let δS_{Dx} be the effective area of the intersections of all microcracks or microcavities which lie in δS;

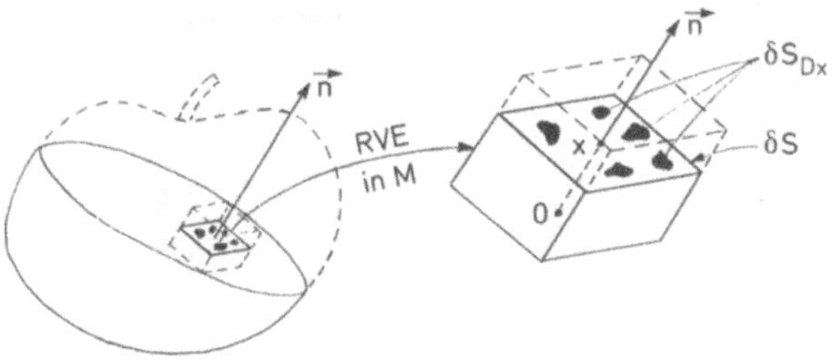

Fig. 1.9. Micro-meso definition of damage (Apple after J. Lemaitre, 1975)

— The value of the damage $D(M, \vec{n}, x)$ attached to the point M in the direction $\vec{n}$ and at the abscissa x is:

$$D(M, \vec{n}, x) = \frac{\delta S_{Dx}}{\delta S}.$$

In order to define a continuous variable over the RVE for its deterioration to failure in two parts, one must look at all the planes varying with x and consider that which is most damaged:

$$D_{(M,\vec{n})} = \underset{(x)}{\mathrm{Max}}[D_{(M,\vec{n},x)}].$$

The coordinate x disappears, and:

$$\boxed{D_{(M,\vec{n})} = \frac{\delta S_D}{\delta S}}$$

It follows from this definition that the value of the scalar variable D (which depends upon the point and the direction considered) is bounded by 0 and 1:

$$0 \leq D \leq 1;$$

$D = 0 \rightarrow$ undamaged RVE material;
$D = 1 \rightarrow$ fully broken RVE material in two parts.
In fact, the failure occurs for $D < 1$ through a process of instability (see Sections 1.2.4 and 3.1.3.).
Consideration of the simple one-dimensional case of a homogeneous damage pictured in Figure 1.10 leads to the simple definition of damage as the effective surface density of microdefects:

$$D = \frac{S_D}{S}.$$

1.2.2 Effective Stress Concept (Y.N. Rabotnov, 1968)

If the RVE of Figure 1.10 is loaded by a force $\vec{F} = \vec{n}F$, the usual uniaxial stress is:

$$\sigma = \frac{F}{S}.$$

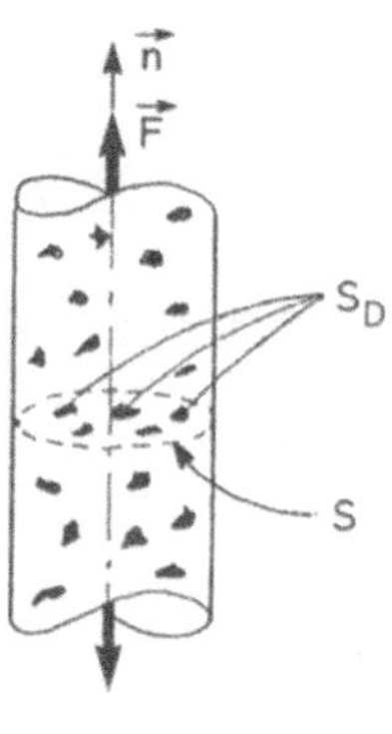

Fig. 1.10. One-dimensional damaged element

If all the defects are open in such a way that no microforces are acting on the surfaces of microcracks or microcavities represented by S_D, it is convenient to introduce an effective stress $\tilde{\sigma}$ related to the surface that effectively resists the load, namely $(S - S_D)$:

$$\tilde{\sigma} = \frac{F}{S - S_D}.$$

Introducing the damage variable $D = \frac{S_D}{S}$,

$$\tilde{\sigma} = \frac{F}{S\left(1 - \frac{S_D}{S}\right)} \quad \text{or} \quad \boxed{\tilde{\sigma} = \frac{\sigma}{1 - D}}$$

This definition is the effective stress on the material in tension. In compression, if some defects close, the damage remaining unchanged, the surface that effectively resists the load is larger than $S - S_D$. In particular, if all the defects close, the effective stress in compression $\tilde{\sigma}^+$ is equal to the usual stress σ. This effect is studied in details in Section 2.3.4.

Coming back to the definition of the ''effective'' area of microcracks δS_D, ''effective'' has to be understood as ''strength'', taking into account the microstress concentrations and mutual interactions of defects loaded in tension or shear. Only micromechanics may give a precise meaning of this concept which will be taken into account globally at the mesoscale through identification of the damage variable by means of its coupling with elasticity or plasticity (see exercise in Section 1.2.5.)

1.2.3 Strain Equivalence Principle (J. Lemaitre, 1971)

A way to avoid a micromechanical analysis for each type of defect and each type of mechanism of damage is to postulate a principle at the mesoscale.

In thermodynamics, the method of local state assumes that the thermomechanical state at a point is completely defined by the time values of a set of continuous state variables depending upon the point considered. This postulate applied at the microscale imposes that the constitutive equations for the strain of a microvolume element are not modified by a neighboring microvolume element containing a microcrack. Extrapolating to the mesoscale, this means that the strain constitutive equations written for the surface $\delta S - \delta S_D$ are not modified by the damage or that the true stress loading on the material is the effective stress $\tilde{\sigma}$ and no longer σ. The following principle results:

''Any strain constitutive equation for a damaged material may be derived in the same way as for a virgin material except that the usual stress is replaced by the effective stress''

Undamaged material	Damaged material
$D = 0$	$0 < D < 1$
$\varepsilon = \mathcal{F}(\sigma, \cdots)$	$\varepsilon = \mathcal{F}\left(\frac{\sigma}{1-D}, \cdots\right)$

↑ Same derivation ↑

This statement is a principle because it has been demonstrat d only in some particular cases of damage through homogenization techniques. It will be applied either to elasticity or plasticity.

1.2.4 Coupling Between Strains and Damage; Rupture Criterion; Damage Threshold

In accordance with the description of the damage mechanics of Sections 1.1.1 and 1.1.2 and as a direct application of the strain equivalence principle, one may write the uniaxial laws of elasticity and plasticity of a damaged material.

Elasticity law

This is direct state coupling through the concept of effective stress:

Undamaged material	Damaged material
$D = 0$	$0 < D < 1$
$\varepsilon_e = \frac{\sigma}{E}$	$\varepsilon_e = \frac{\sigma}{E(1-D)}$

Contraction for isotropic damage:

$$\varepsilon_{22}^e = \varepsilon_{33}^e = -\nu\varepsilon_e;$$

E is Young's modulus of the undamaged material;
ν is Poisson's ratio.
The elasticity modulus of the damaged material defined by the ratio $\tilde{E} = \frac{\sigma}{\varepsilon_e}$ is

$$\tilde{E} = E(1 - D).$$

Figure 1.11 shows an example of the variation of the elasticity modulus as ductile damage progresses. It can also be evaluated from the microcrack pattern by means of micromechanics (B. Budiansky, R. O'Connell, 1976).

Plasticity

This is a kinetic coupling on the evolution of plastic strain which has to be written in the plasticity criterion used to derive the kinetic constitutive equations.
In order to model plasticity two kinds of strain hardening are usually considered:

- the isotropic hardening related to the density of dislocations or flow arrests;
- the kinematic hardening related to the state of internal microstress concentrations. The corresponding back stress defines the center of the elastic domain in tension compression (or in three dimensions).

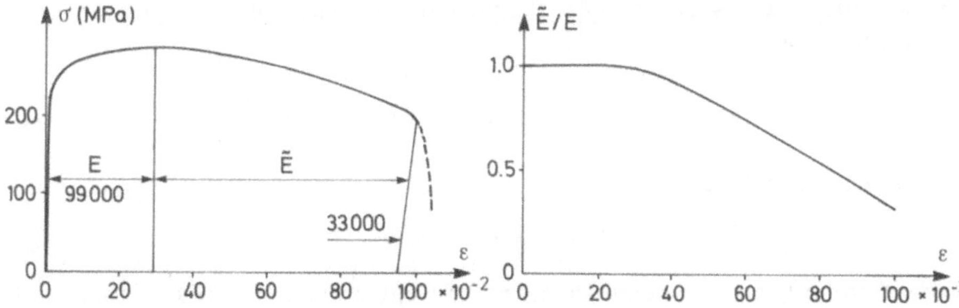

Fig. 1.11. Variation of the elasticity modulus with damage for copper (99.9%) (after J. Dufailly, 1975)

If σ_y is the yield stress, R the stress due to isotropic hardening and X the back stress, both functions of the plastic strain, the one-dimensional plasticity criterion defining the current threshold of yield limit, and represented in Figure 1.12 is

$$\sigma = \sigma_y + R + X$$
$$\text{or } f = |\sigma - X| - R - \sigma_y = 0.$$

f is the yield function from which the kinetic constitutive equation for plastic strain is derived (see Section 2.1.3):

$$\exists \dot{\varepsilon}_p \neq 0 \text{ if } \begin{cases} f = 0 \\ \text{and} \\ \dot{f} = 0 \end{cases}, \qquad \dot{\varepsilon}_p = 0 \text{ if } \begin{cases} f < 0 \\ \text{or} \\ \dot{f} < 0 \end{cases}$$

To achieve this, let us write the total strain as

$$\varepsilon = \varepsilon_e + \varepsilon_p.$$

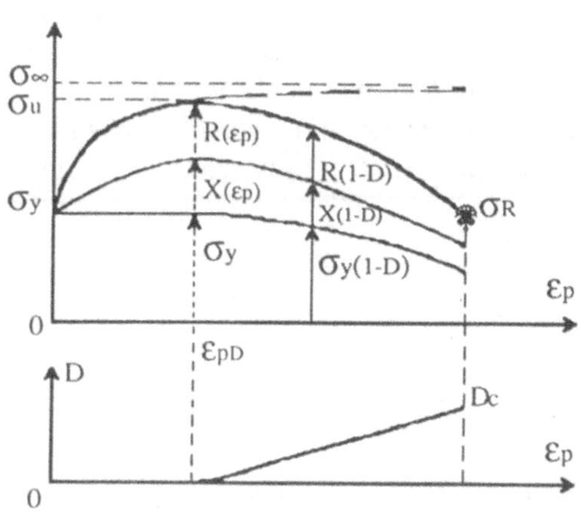

Fig. 1.12. Schematic evolution of the plasticity yield criterion with damage

When damage occurs, according to the principle of equivalence, the yield function f must be written as:

$$f = \left| \frac{\sigma}{1-D} - X \right| - R - \sigma_y = 0$$

Experiments and the equation

$$\sigma = (\sigma_y + R + X)(1 - D)$$

show that the damage equally decreases the yield stress, the isotropic strain hardening stress and the back stress as shown in Figure 1.12.

Rupture criterion

The rupture at the mesoscale is a crack initiation which occupies the whole surface of the RVE; that is, $D = 1$. In many cases this is caused by a process of instability which suddenly induces the decohesion of atoms in the remaining resisting area. It corresponds to a critical value of damage D_c, which depends upon the material and the conditions of loading.

The final decohesion of atoms is characterized by a critical value of the effective stress acting on the resisting area. Let us call it σ_∞ because it is the maximum stress that could be applied to the material (see Figure 1.12):

$$\tilde{\sigma} = \frac{\sigma}{1 - D_c} = \sigma_\infty$$

Practically σ_∞ may be approximated by the ultimate stress σ_u, which is easier to identify but is always smaller. Then

$$D_c \simeq 1 - \frac{\sigma}{\sigma_u}$$

gives the critical value of the damage at a mesocrack initiation occurring for the unidimensional stress σ. The ultimate stress σ_u being identified as a material characteristic, D_c may vary between $D_c \simeq 0$ for pure brittle fracture to $D_c \simeq 1$ for pure ductile fracture but usually D_c remains of the order of 0.2 to 0.5.

This relation, applied to the pure monotonic tension test, which is taken as a reference, defines the corresponding critical damage D_{1c} considered as a material characteristic:

$$D_{1c} = 1 - \frac{\sigma_R}{\sigma_u}$$

where σ_R is the stress to rupture.

Damage threshold

Before the microcracks are initiated, creating the damage modelled by D, they must nucleate by the accumulation of microstresses accompanying incompatibilities of microstrains or by the accumulation of dislocations in metals. This corresponds in the pure tension case to a certain value ε_{pD} of the plastic strain below which no damage by microcracking occurs:

$$\varepsilon_p < \varepsilon_{pD} \rightarrow D = 0.$$

Finally, the four main relations which comprise the basis of damage mechanics are:

$$\varepsilon_e = \frac{\sigma}{E(1-D)} \text{ for elasticity;}$$

$$\left|\frac{\sigma}{1-D} - X\right| - R - \sigma_y = 0 \text{ as the plastic yield criterion;}$$

$$\varepsilon_p < \varepsilon_{pD} \rightarrow D = 0 \text{ as the damage threshold;}$$

$$D = Dc \rightarrow \text{ crack initiation.}$$

1.2.5 Exercise on the Micromechanics of the Effective Damage Area

In Section 1.2.1., the one-dimensional variable of damage has been defined as the ''effective'' surface density of intersections of microcracks and cavities in the most damaged section. Let us illustrate the meaning of this, using a particular example. Consider a three-dimensional mesovolume element containing a single circular crack loaded in mode I by a load F perpendicular to the plane of the crack (Figure 1.13.).

The overall damage defined by the continuum damage theory is the damage in the plane of the crack,

$$D = \frac{\delta S_D}{\delta S};$$

namely,

$$D = \frac{\pi a^2}{l^2} k,$$

where k is the correction factor due to the stress concentration in the vicinity of the crack tip line to define the effective damaged area.

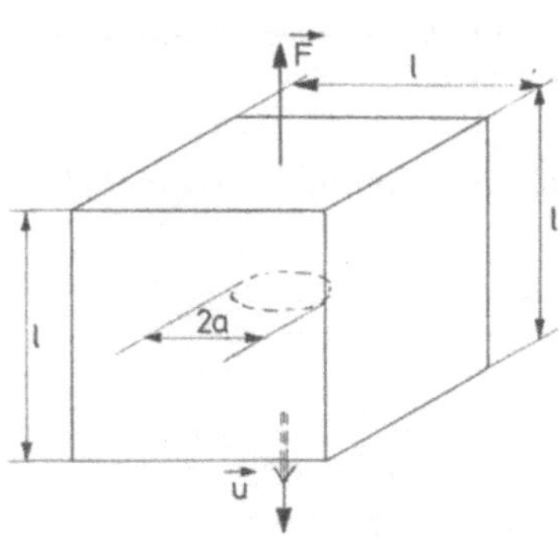

Fig. 1.13. Single cracked meso volume element

Calculation of k as a function of a/l

Applying the strain equivalence principle together with the effective stress concept, the damage may also be defined as the variation of the stiffness R_a of the mesoelement

$$\vec{F} = (1-D)R_0\vec{u},$$

where $\vec{u}$ is the elastic displacement corresponding to the load $\vec{F}$, and R_0 the initial stiffness of the uncracked element:

$$R_a = R_0(1-D),$$

with

$$R_0 = \frac{El^2}{l} = El,$$

E being the Young's modulus of the material.

Calculation of R_a from the two concepts of classical fracture mechanics

- the strain energy release rate,

W being the strain energy and A the crack area:

$$G = -\frac{1}{2}\frac{\partial W}{\partial A}\bigg|_{F=\text{const}} = -\frac{1}{2}\frac{F^2}{R_a^2}\frac{dR_a}{2\pi a da};$$

- and the stress intensity factor for a circular crack in an infinite body loaded in mode I:

$$K = \frac{2}{\pi}\sigma_\infty\sqrt{\pi a} = \frac{2}{\pi}\frac{F}{l^2}\sqrt{\pi a}.$$

For a case of generalized plane strain (ν being the Poisson's ratio),

$$G = \frac{K^2}{E}(1-\nu^2).$$

Writing the equality of the two expressions of G gives a differential equation,

$$-\frac{l^4}{16}\frac{dR_a}{R_a^2} = \frac{1-\nu^2}{E}a^2 da,$$

which can be integrated from $a = 0 \to R_a = R_0$, to the current state (a, R_a). The result is

$$R_a = \left[\frac{1}{R_0} + \frac{16(1-\nu^2)a^3}{3El^4}\right]^{-1}.$$

Assuming a small crack $(a/l \ll 1)$ giving a small variation of the stiffness,

$$R_a = R_0\left(1 - \frac{16(1-\nu^2)a^3}{3El^4}R_0\right).$$

This gives an expression for D:

$$D = 1 - \frac{R_a}{R_0} = \frac{16(1-\nu^2)a^3}{3l^3} = \frac{\pi a^2}{l^2}\frac{16(1-\nu^2)}{3\pi}\frac{a}{l},$$

from which it follows that

$$k = \frac{16(1-\nu^2)}{3\pi}\frac{a}{l} \simeq 1.55\frac{a}{l}.$$

This means that for this particular case and for $a/l \ll 1$, the value of the damage is always smaller than the crack density.

Determining the effective stress

In this simple example where the damage is D, the effective stress can obviously be calculated as in tension when the crack is open;

$$\tilde{\sigma} = \frac{\sigma}{1-D} = \frac{F^+}{l^2(1-D)} \quad \text{in tension when the crack is open;}$$

$$\tilde{\sigma} = \sigma = \frac{F^-}{l^2} \quad \text{in compression when the crack is closed.}$$

This crack closure effect will be studied in Section 2.3.4.

1.3 Measurement of Damage

1.3.1 Direct Measurements

Returning to the definition of the damage as $D = \delta S_D/\delta S$, the direct measurement consists in the evaluation of the total crack areas δS_D lying on a surface δS at mesoscale. Assuming the correction factor k to be equal to 1, this can be done by observing micrograph pictures. To observe a picture of an approximately (100-mm)^2 section of a RVE, a maximum magnification of 1000 is sufficient for metals, and 1 to 10 is enough for concrete. This is of course a destructive method and tedious to practice. An example is given in Figure 1.14 showing ductile microcracks and dimple formations in an high-alloy steel:

At the scale of real observation:

δS_D = 1163 mm^2

δS = 11317 mm^2

In the plane of observation:

$$D = \frac{\delta S_D}{\delta S} \simeq 0.1$$

If the damage consists of microcracks which are not perfectly homogeneous in the RVE, there is almost no chance of obtaining even one crack in any plane of intersection! Therefore the damage can only be evaluated by means of the lines which indicate the intersection of the microcracks with the plane of observation. An equivalent isotropic damage may be obtained as follows : Consider a micromechanical element of n cells of size $d \times d \times d$, each containing a crack of different apparent size a_i (Figure 1.15). Assuming square cracks for simplicity, and a correction factor equal to 1, the damage in the plane of the crack of each cell i is

$$D_i = \frac{\delta S_D}{\sigma S} \simeq \frac{a_i^2}{d^2}.$$

Fig. 1.14. Direct evaluation of damage (after Engel, Klingele and C. Lienard)

The equivalent isotropic damage in the plane of observation of the n cells is taken as the mean value of the damage in each cell. This is the simplest way to homogenize discrete quantities in order to define a continuous variable

$$D = \frac{\sum_{i=1}^{n} \frac{a_i^2}{d^2}}{n} = \frac{\sum_{i=1}^{n} a_i^2}{nd^2}$$

If a micrograph of size l^2 is considered with n cells corresponding to n cracks, the surface of the micrographic may be expressed as

$$l^2 = n \cdot d^2.$$

The value of the damage is simply:

$$D = \frac{\Sigma a_i^2}{l^2}.$$

An example of damage is shown in Figure 1.15., where grain boundary precipitates of titanium carbide caused intercrystalline failure in a martensitic NiCoMo steel subjected to triaxial stresses.

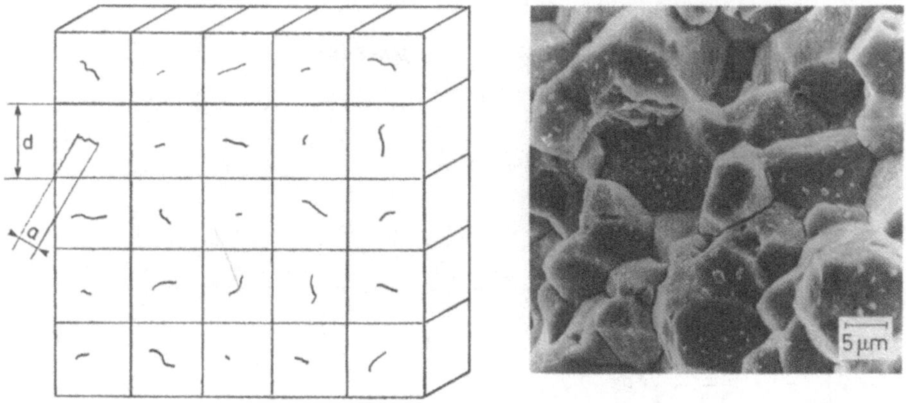

Fig. 1.15. Micromechanic model and direct evaluation of damage (after L. Engel, H. Klingele and C. Lienard)

At the scale of real observation:
$\Sigma a_i^2 = 9000\ \text{mm}^2$
$l^2 = 10200\ \text{mm}^2$
In the plane of observation

$$D = \frac{\Sigma a_i^2}{l^2} \simeq 0.88$$

1.3.2 Variation of the Elasticity Modulus

This is a nondirect measurement based on the influence of damage on elasticity through the state coupling as explained in Section 1.2.4:

$$\varepsilon_e = \frac{\sigma}{E(1-D)}.$$

This destructive method requires the machining of specimens in order to run mechanical tests. It assumes uniform homogeneous damage in the specimen gauge section.

If $\tilde{E} = E(1-D)$ is considered as the effective elasticity modulus of the damaged material, the values of the damage may be derived from measurements of $\tilde{E}$, provided that Young's modulus E is known:

$$\boxed{D = 1 - \frac{\tilde{E}}{E}}$$

This very useful method requires accurate strain measurements. Strain gauges are commonly used and $\tilde{E}$ is most accurately measured during unloadings. An example

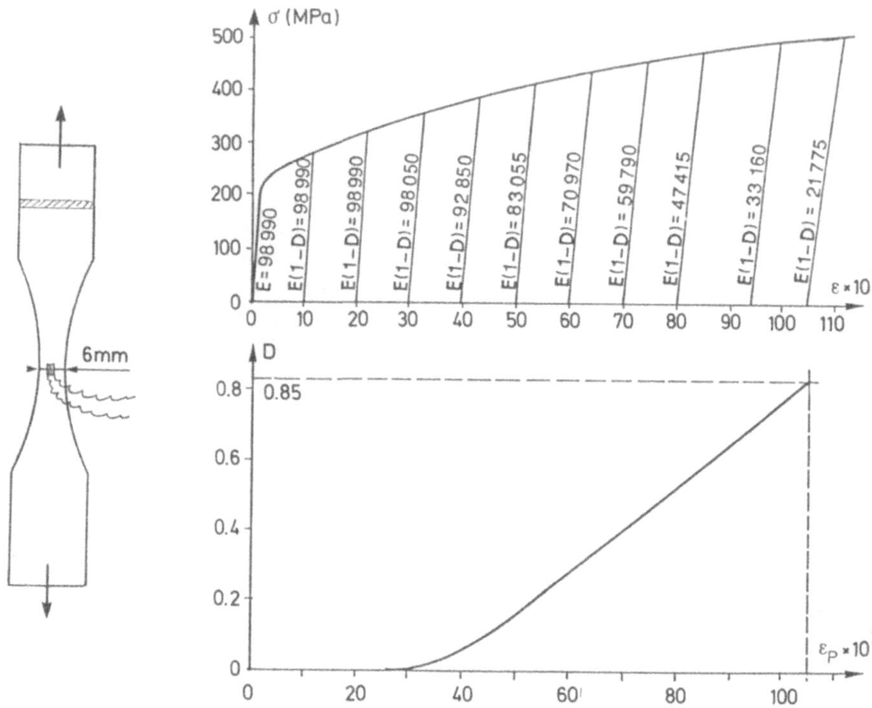

Fig. 1.16. Measurement of ductile damage on 99.9% copper at room temperature (after J. Dufailly)

is given in Figure 1.16 for ductile damage of copper induced by large deformations, where σ_v is the true stress $\sigma_v = \sigma(1+\varepsilon)$

This technique may be used for any kind of damage as long as the damage is uniformly distributed in the volume on which the strain is measured which is the main limitation of the method. If the damage is too greatly localized, as for high cycle fatigue of metals, for example, another method must be used.

Some other precautions related to non-linearities have to be taken. At the beginning and at the end of the unloading paths in the plane (σ, ε) there are small nonlinearities, owing to viscous or hardening effects and also due to the experimental devices as well. It is best to ignore them and to identify $\tilde{E}$ in the range

$$0.15\frac{F_{\max}}{S} < \frac{F}{S} < 0.85\frac{F_{\max}}{S}$$

It is most important to always use the same procedure to evaluate E and the evolution of $\tilde{E}$.

For ductile or low cycle fatigue damage in metals, the procedure could be somewhat disturbed by an early decrease of $\tilde{E}$ at low strain levels or during the first cycles. This is due to microplasticity related to reversible movements of dislocations, and to texture development, but not to damage. As this phenomenon is rapidly saturated, it is easy to consider:

$$D = 0 \text{ for } \varepsilon < \varepsilon_D \text{ (the damage strain threshold)}$$

or

$$D = 0 \text{ for } N < N^* \text{ (the number of cycles to stabilization).}$$

For polymers or composites, to avoid the viscous effect of viscoelasticity, the strain rate during unloadings must be the same for the measurement of E and $\tilde{E}$. An example of a damage measurement on a carbon-carbon composite is given in Figure 1.17.

For concrete it is important to check the "uniform" distribution of cracks, as the method is no longer valid if a single big crack develops. An example is given in Figure 1.18.

Ultrasonic waves propagation

Another technique for evaluating damage based on the variation of the elasticity modulus consists in measuring the speed of ultrasonic waves.

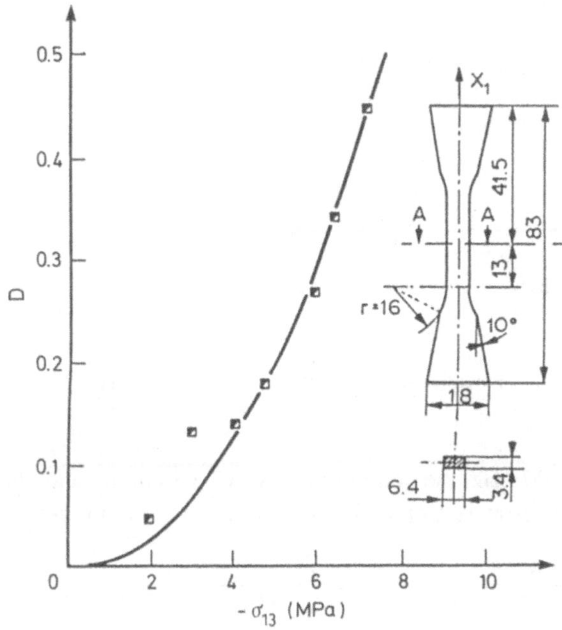

Fig. 1.17. Evolution of brittle damage in a shear direction of a three-dimensional carbon-carbon composite (after M. Poss)

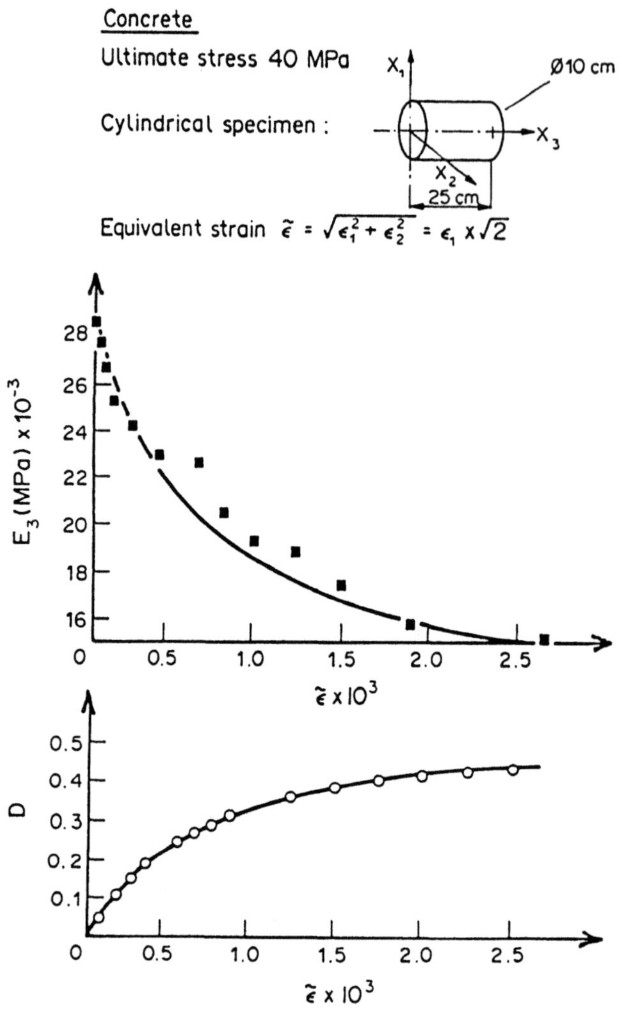

Fig. 1.18. Evolution of brittle damage in compression for a concrete (σ_u = 40 MPa) (after J. Mazars)

For frequencies higher than 200 kHz the longitudinal wave speed v_L and the transversal wave speed v_T in a linear isotropic elastic cylindrical medium are:

$$v_L^2 = \frac{E}{\rho} \frac{1-\nu}{(1+\nu)(1-2\nu)} \qquad v_T^2 = \frac{E}{\rho} \frac{1}{2(1+\nu)},$$

where E is the Young's modulus, ρ the density and ν Poisson's ratio. A measure of the longitudinal wave speed of a damaged material gives

$$\tilde{v}_L^2 = \frac{\tilde{E}}{\tilde{\rho}} \frac{1-\nu}{(1+\nu)(1-2\nu)},$$

where $\tilde{E}$ and $\tilde{\rho}$ are the actual damaged elastic modulus and density. Poisson's ratio does not vary with damage if elasticity is isotropic, and damage a scalar.
The damage is calculated by:

$$D = 1 - \frac{\tilde{E}}{E} = 1 - \frac{\tilde{\rho}\,\tilde{v}_L^2}{\rho\, v_L^2}.$$

If the damage consists mainly of microcracks or if small cavitation is considered, $(\tilde{\rho}/\rho) \simeq 1$ and:

$$D \simeq 1 - \frac{\tilde{v}_L^2}{v_L^2}.$$

The method considered here is destructive: indeed, to measure the speed v_L or the time a wave takes to propagate through a certain thickness, one needs to limit that thickness by two surfaces. If the space distribution of the damage is not uniform, the thickness must be of an order of magnitude coherent with that of the RVE, therefore that requires the body to being analyzed to be cut into parts.
The present limitations of this method are related to the size, to be analyzed, which, for metals, is too small in comparison with ultrasonic transducers size and accuracy of time measurements. Nevertheless the method gives good results for concrete.
The ranges of frequency to be used are:

1.0 to 50 MHz for metals,
0.1 to 5 MHz for polymers,
0.1 to 5 MHz for woods,
0.1 to 1 MHz for concrete.

An example is given in Figure 1.19.
This method can be improved so as to be used in situ without causing any destruction by working with Rayleigh's surface waves and attenuating of the ultrasonic signal.

1.3.3 Variation of the Microhardness

This is another nondirect measurement based on the second fundamental equation of the conclusion of Section 1.2.4: the influence of damage on the plasticity-yield criterion through the kinetic coupling. Assume an equivalent unidimensional state of stress, it is written

$$\left|\frac{\sigma}{1-D} - X\right| - R - \sigma_y = 0$$

As microhardness is a process of very small indentation, this method may be considered as practically nondestructive. The test consists in inserting a diamond indenter in the material, the hardness H being defined by the mean stress in space:

$$H = \sigma = \frac{F}{S}.$$

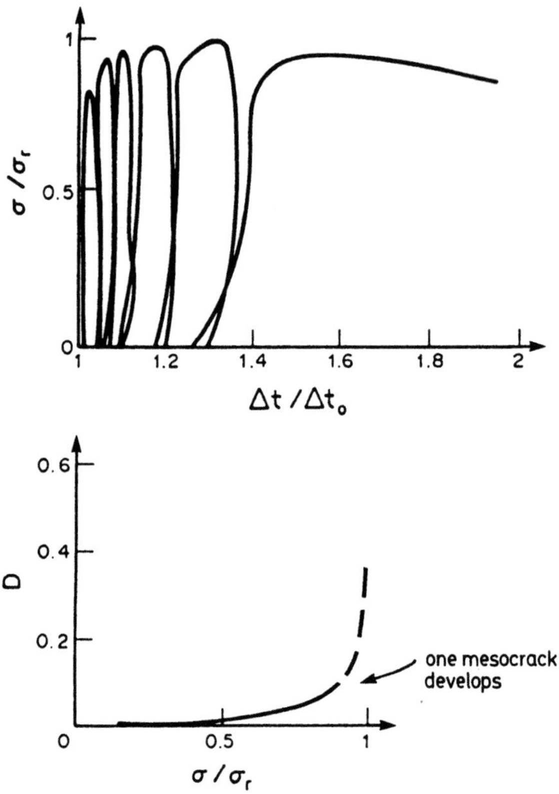

Fig. 1.19. Evolution of brittle damage from the propagation time Δt of ultrasonic waves in concrete (after Y. Berthaud).

The load F on the indenter is chosen so as to obtain a projected indented area S of the same order of magnitude as that of the RVE.
Theoretical analyses and many experimental results prove a linear relationship between H and the plasticity threshold σ_s, k' being the coefficient of proportionality:

$$H = k'\sigma_s .$$

This threshold corresponds to the actual yield stress,

$$\sigma_s = (\sigma_y + R + X)(1 - D).$$

then

$$H = k'(\sigma_y + R + X)(1 - D).$$

In fact, the microhardness test itself increases the strain hardening by an amount which corresponds to a plastic strain ε_p^H of the order of 5 to 8%. H is then always related to $(\varepsilon_p + \varepsilon_p^H)$, ε_p being the current plastic strain.

If $H^* = k'(\sigma_y + R + X)$ is the microhardness of the material which would exist without any damage for $(\varepsilon_p + \varepsilon_p^H)$, and $H = k'(\sigma_y + R + X)(1 - D)$ the actual micro-hardness for $(\varepsilon_p + \varepsilon_p^H)$, then

$$\boxed{D = 1 - \frac{H}{H^*}}.$$

H is measured, and H^* has to be evaluated as proposed below:

— For high cycle fatigue, damage occurs while the stress remains below the yield stress so that $\varepsilon_p \simeq 0$ and

$$H^*_{(\varepsilon_p^H)} = k'\sigma_{s(\varepsilon_p^H)}.$$

H^* may be measured on a nondamaged part of the material.

— For the case of low cycle fatigue, we may consider that the strain hardening is saturated so that $R = R_\infty$ and $X = X_\infty$:

$$H^*_{(\varepsilon_p + \varepsilon_p^H)} = k'(\sigma_y + R_\infty + X_\infty) = \text{const.}$$

H^* is obtained by a measurement on the fully strain-hardened but not yet damaged material.

— For the case of ductile damage, damage and strain hardening occur simultaneously and H^* has to be obtained by some extrapolation procedure.

Figure 1.20. gives a comparison of results obtained by measuring the evolution of ductile damage and those obtained by measuring the variation of the elastic modulus.

Apart from the quasi-non destructive character of this method, it is also a convernient way to evaluate damage fields if the microhardness test is automatized (Figure 1.21.). Automatization makes it possible to measure the state of damage of a structure submitted to some accident which, for safety purposes can be introduced as an initial condition in the calculation of the residual strength (see Section 4.4.1.).

Another possibility of this method of measurement is to deduce the value of the strain hardening that may exist in some part of the structure after loading.

Let us call H_y the microhardness measured at a point where the material was certainly loaded below the yield stress σ_y:

$$H_y = k'\sigma_y,$$

and let H be the microhardness at the point where the strain hardening has to be evaluated:

$$H = k'(\sigma_y + R + X)(1 - D)$$

or

$$H = \frac{H_y}{\sigma_y}(\sigma_y + R + X)(1 - D);$$

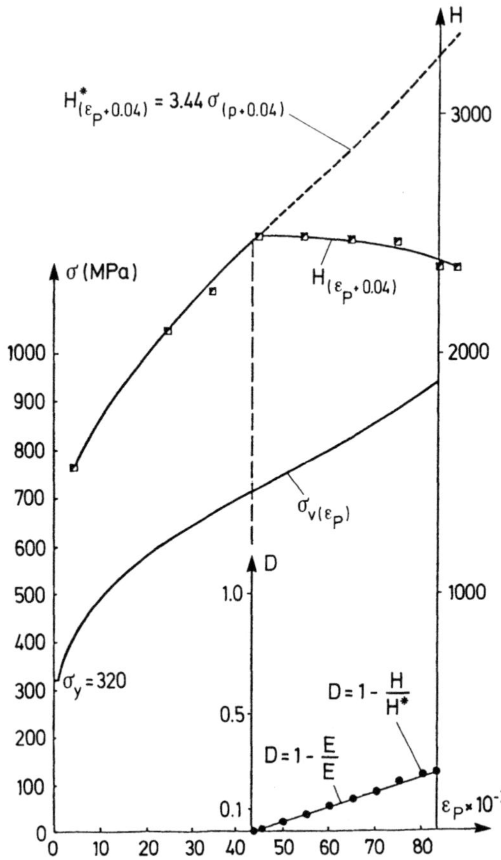

Fig. 1.20. Evolution of ductile damage by microhardness on AISI 1010 at room temperature (after R. Billardon)

then

$$R+X = \sigma_y\left(\frac{H}{H_y(1-D)}-1\right)$$

or

$$R+X = \sigma_y\left(\frac{H^*}{H_y}-1\right),$$

H^* having the same meaning as previously. $(R+X)$ is the sum of the isotropic strain-hardening stress and the kinematic back stress of an uniaxial stress equivalent to the three-dimensional state of stress produced by the indentation. If no kinematic hardening is assumed, R may be introduced easily as an initial value, for example, in a three-dimensional analysis as explained in Section 4.4.1.

1.3.4 Other Methods

Several other methods, which are based upon the influence of damage on some physical or mechanically measurable properties, may also be used.

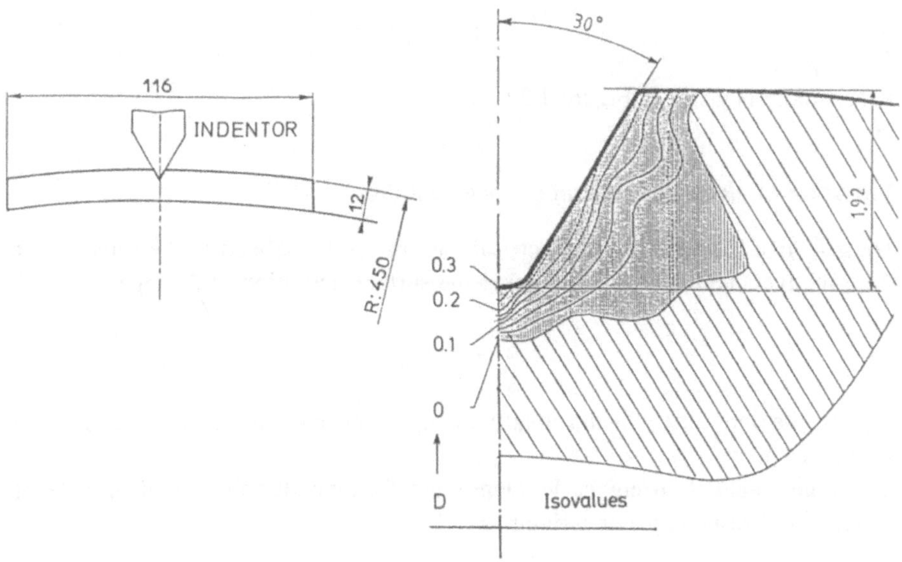

Fig. 1.21. Damage field on a plastic deformed built notched specimen as obtained from microhardness measurements on AISI 1010 (after J. Dufailly, R. Billardon)

Variation of density

In the case of pure ductile damage, the defects are cavities which can be assumed to be roughly spherical; this means that the volume increases with damage. The corresponding decrease of density is measurable with apparatuses based on the Archimedean principle.

If $(\tilde{\rho}-\rho)/\rho$ is the relative variation of density between the damaged state $\tilde{\rho}$ and the initial nondamaged state ρ, it easy, by means of micromechanics, considering a spherical cavity of radius r in a spherical R.V.E.of initial radius R and mass m, to derive the following relationships between the surface damage D and the variation of density or porosity, assuming no residual microstress:

$$\rho = \frac{m}{\frac{4}{3}\pi R^3},$$

$$\tilde{\rho} = \frac{m}{\frac{4}{3}\pi (R^3 + r^3)},$$

$$\frac{\tilde{\rho}-\rho}{\rho} = \frac{R^3}{R^3+r^3} - 1 = \frac{-r^3}{R^3+r^3},$$

$$D = \frac{\delta S_D}{\delta S} = \frac{\pi r^2}{\pi (R^3+r^3)^{2/3}} = \left(\frac{r^3}{R^3+r^3}\right)^{2/3},$$

$$D = \left(1 - \frac{\tilde{\rho}}{\rho}\right)^{2/3}.$$

An example is given in Figure 1.22.

Variation of electrical resistance (potential drop)

The effective intensity of the electrical current can be defined in the same way as the effective stress was defined, using the surface definition of damage:

$$\tilde{i} = \frac{i}{1-D},$$

$\tilde{i}$ being the intensity which actually exists in the cohesive parts of a damaged volume element.

Given the potential difference V, Ohm's law for a nondamaged element of length l, area s and resistivity r is written as

$$V = r\frac{l}{s}i;$$

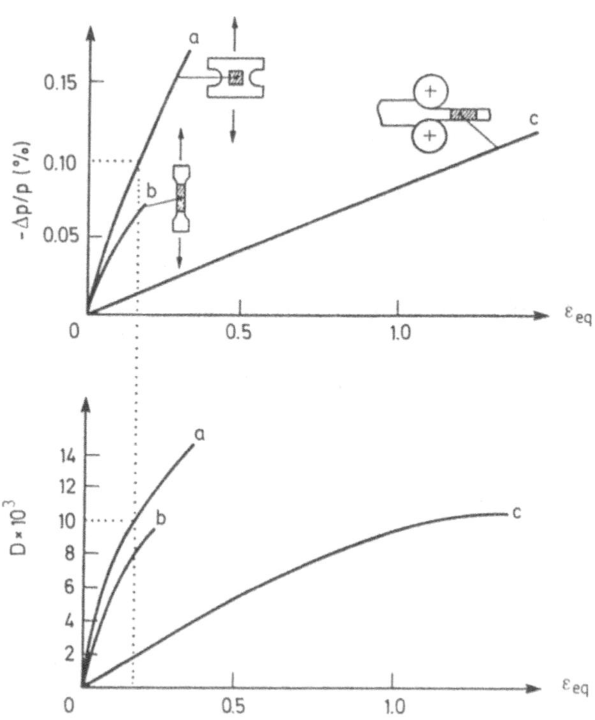

Fig. 1.22. Evauation of ductile damage on steel for different strain hardenings (after F. Moussy)

whereas, for a damaged element of the same size,

$$\tilde{V} = \tilde{r}\frac{l}{s}\tilde{i},$$

where $\tilde{r}$ is the resistivity affected by the damage by means of the change of volume only $(\tilde{\rho}-\rho)/\rho$ (ρ being the density). Bridgman's law gives $\tilde{r}$ as:

$$\tilde{r} = r\left(1+K\frac{\rho-\tilde{\rho}}{\rho}\right) = r(1+KD^{3/2}).$$

K is a coefficient which is approximately 2 for metals.
If the same intensity i is considered for the nondamaged and the damaged elements, the damage D may be derived from the two expressions

$$\frac{V}{\tilde{V}} = \frac{r}{\tilde{r}}\frac{\frac{l}{s}i}{\frac{l}{s}\frac{i}{1-D}},$$

$$D = 1-\frac{V}{\tilde{V}}\frac{\tilde{r}}{r} \text{ with } \frac{\tilde{r}}{r} = 1+KD^{3/2}.$$

For small values of D, the correction term $(\tilde{r}/r)$ due to the volume change is close to 1 (for instance, $D \simeq 0.1 \rightarrow \tilde{r}/r = 1.064$); then

$$D \simeq 1-\frac{V}{\tilde{V}}.$$

This method is known as the "potential drop method".
Some comparisons between measurements of damage obtained with this method and from elasticity modulus variations $D = 1-\tilde{E}/E$ or methods described hereafter are shown in Figure 1.23.
As the damage measurements are made during tension tests, a correction of these examples is required to account for the length change $l(1+\epsilon)$, the area change $s(1-2\nu\varepsilon_e-\varepsilon_p)$ and the volume change due to elastic strain: $sl(1-2\nu)\varepsilon_e$. The agreement between the results obtained by the different methods is sufficient for the practical accuracy involved.

Variation of the cyclic plasticity response (stress amplitude drop)

The influence of damage on plasticity may be used to measure the low cycle fatigue damage.
The one-dimensional law of cyclic plasticity at stabilization may be written as a power relationship between the amplitude of stress $\Delta\sigma$ and the amplitude of plastic strain $\Delta\varepsilon_p$ (see Section 3.4.1.):

$$\Delta\varepsilon_p = \left(\frac{\Delta\sigma}{K_p}\right)^M$$

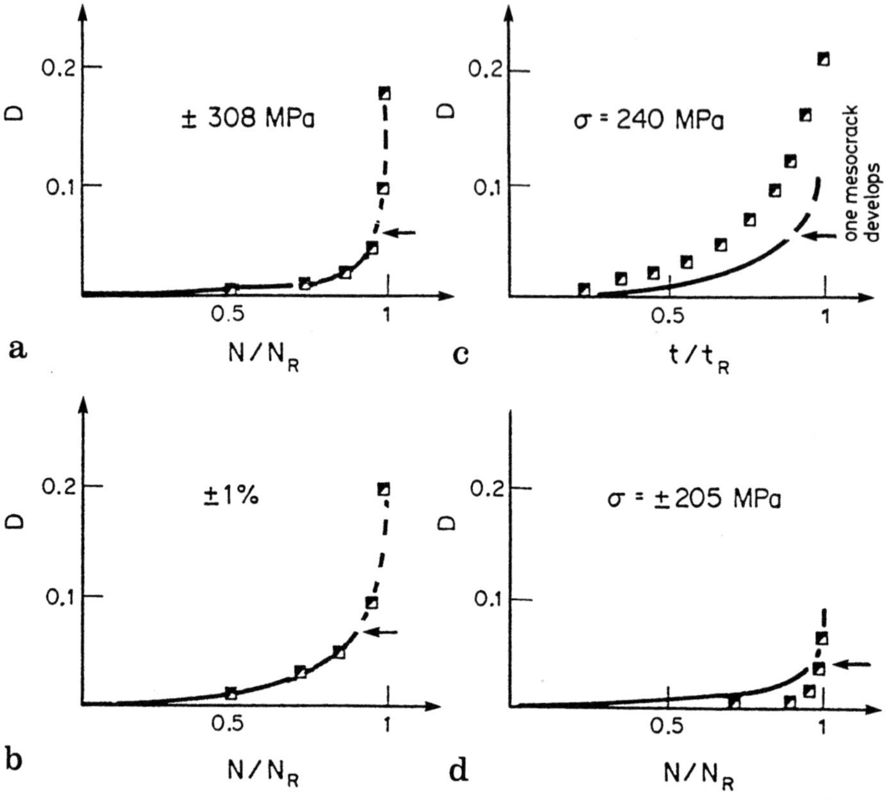

Fig. 1.23. Comparison between damage measurements by variation of the electrical potential and by **a** the elasticity modulus change on stress-controlled fatigue of A 316 stainless steel, **b** the elasticity modulus change on strain-controlled fatigue of A 316 stainless steel, **c** the tertiary creep method on creep damage of IN 100 superalloy at 1000°C, **d** the cyclic plasticity response on stress controlled fatigue of IN 100 superalloy at 1000°C (after G. Cailletaud)

for a nondamaged material, and

$$\Delta\varepsilon_p = \left[\frac{\Delta\sigma}{K_p(1-D)}\right]^M$$

for a damaged material, in application of the principle of strain equivalence, where K_p and M are material parameters.

Considering a test at constant plastic strain amplitude, if $\Delta\sigma^*$ is the stress amplitude at stabilization at the end of the cyclic softening or hardening period and before the beginning of the damage process:

$$\Delta\varepsilon_p = \left(\frac{\Delta\sigma^*}{K_p}\right)^M = \left[\frac{\Delta\sigma}{K_p(1-D)}\right]^M ,$$

from which it follows that

$$D = 1 - \frac{\Delta\sigma}{\Delta\sigma^*}.$$

Another way to derive this expression is proposed as an exercise in Section 1.3.5. This method successfully identifies the evolution of damage during low cycle fatigue of metals except when stabilization of the cyclic softening or hardening does not occur. An example is given in Figure 1.24.

Tertiary creep response

Creep damage occurs in metals loaded at temperatures above approximately one-third of the melting temperature. To identify creep damage during creep tests (constant σ) it is convenient to use once more the principle of strain equivalence applied here to Norton's law of secondary creep (see Section 2.1.3.).

$$\dot{\varepsilon}_p^* = \left(\frac{\sigma}{K_v}\right)^N,$$

K_v and N being temperature-dependent material parameters.

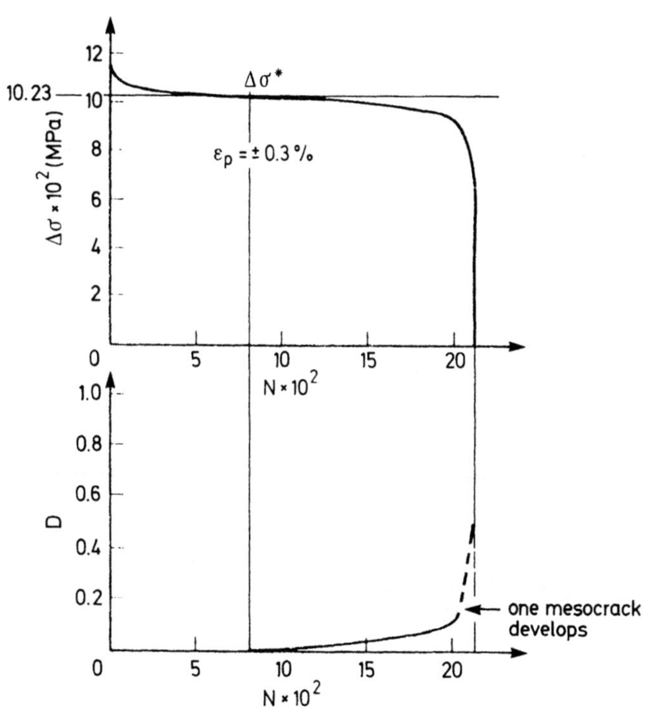

Fig. 1.24. Evolution of low cycle fatigue damage on AISI 1010 deep drawn steel

Assuming that the damage process begins at the end of secondary creep, during tertiary creep one may write:

$$\dot{\varepsilon}_p = \left[\frac{\sigma}{K_v(1-D)}\right]^N ,$$

from which one derives

$$D = 1 - \left(\frac{\dot{\varepsilon}_p^*}{\dot{\varepsilon}_p}\right)^{1/N} ,$$

where $\dot{\varepsilon}_p^*$ is the minimum creep rate.

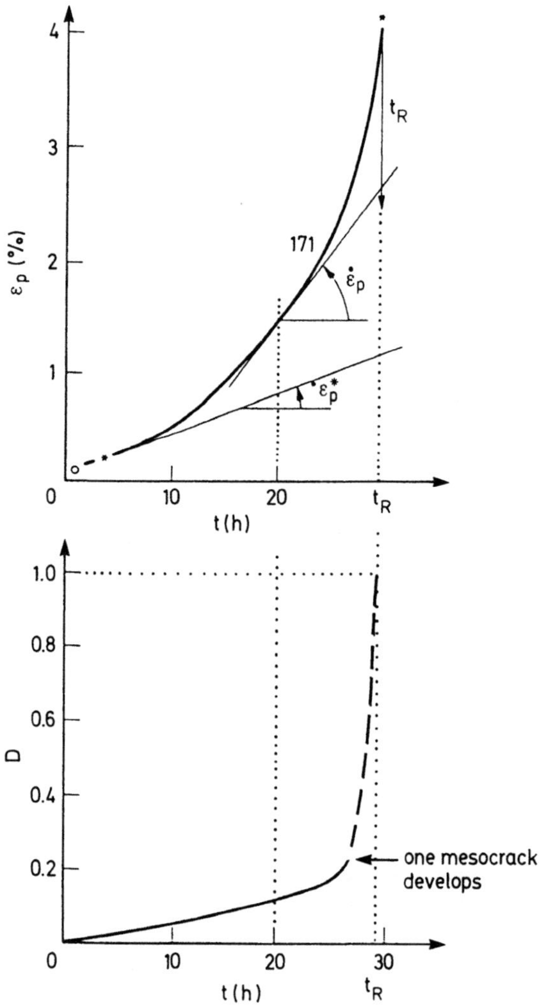

Fig. 1.25. Evaluation of creep damage on IN 100 superalloy at 1000°C (after H. Policella)

This method yields good results which are in accordance with those obtained by measuring the variation of the elasticity modulus. An example is given in Figure 1.25.

Acoustic emission

It is a good method for detecting the location of the damaged zone, but the results remain qualitative as far as the values of the variable D are concerned.
By way of conclusion, Figure 1.26 offers, in the form of a wine selection chart (!), some advice for choosing the proper method of damage measurement, depending on the kind of damage involved.

	Damage	Brittle	Ductile	Creep	Low cycle fatigue	High cycle fatigue
Micrography	$D = \frac{\partial S_D}{\partial S}$	*	**	**	*	*
Density	$D = \left(1 - \frac{\tilde{\rho}}{\rho}\right)^{2/3}$		**	*	*	
Elasticity modulus	$D = 1 - \frac{\tilde{E}}{E}$	**	***	***	***	
Ultrasonic waves	$D = 1 - \frac{\tilde{V}_L^2}{V_L^2}$	***	**	**	*	*
Cyclic stress amplitude	$D = 1 - \frac{\Delta\sigma}{\Delta\sigma^*}$		*	*	**	*
Tertiary creep	$D = 1 - \left(\frac{\dot{\varepsilon}_p^*}{\dot{\varepsilon}_p}\right)^{1/N}$		*	***	*	
Micro-hardness	$D = 1 - \frac{H}{H^*}$	**	***	**	***	*
Electrical resistance	$D = 1 - \frac{V}{\tilde{V}}$	*	**	**	*	*

three stars	***	means "very good"	like "Bourgogne Chambertin 1985"
two stars	**	means "good"	like "Bourgogne Pommard 1983"
one stars	*	means "try to see"	like "Beaujolais 1989"
no stars		means "do not try"	like ???

A votre santé!

Fig. 1.26. Quality chart of methods of damage measurement

1.3.5 Exercise on Measurement of Damage by the Stress Amplitude Drop

Consider a cyclic strain-controlled test in tension compression where the drop of the stress range caused by low cycle fatigue damage after saturation of the strain hardening is recorded as a function of the number of cycles (Figure 1.27). The amplitude of strain is such that the maximum stress reaches a value close to the ultimate stress σ_u at saturation.

From the one-dimensional coupled strain-damage constitutive equations of Section 1.2.4, *demonstrate that the elastic strain amplitude remains constant*:

$$\varepsilon_e = \frac{\sigma}{E(1-D)},$$

$$\frac{\sigma}{(1-D)} - X - R - \sigma_y = 0.$$

When the strain hardening is saturated and the damage still zero,

$$X + R + \sigma_y = \sigma_\infty \simeq \sigma_u;$$

then $\dfrac{\sigma_M}{1-D} = \sigma_u$ and $\varepsilon_{eM} = \dfrac{\sigma_u}{E}$ = const.

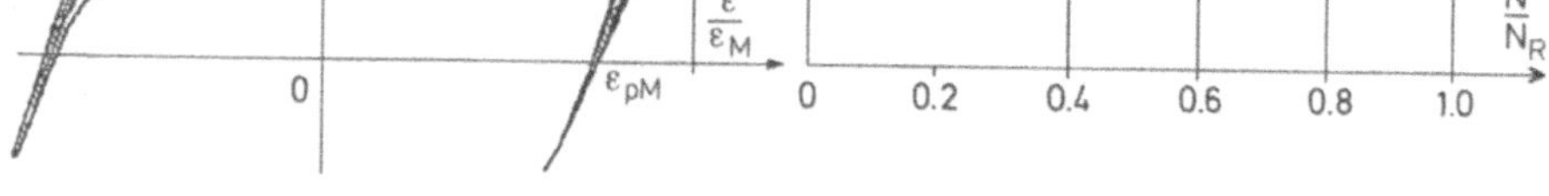

Fig. 1.27. Cyclic strain controlled test on MAR M 509 stainless steel.

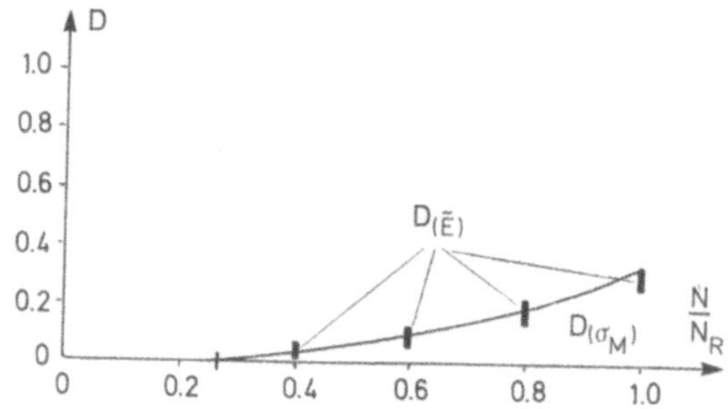

Fig. 1.28. Evolution of the damage measured by the stress amplitude drop and by the elasticity modulus change

From the previous equation,

$$D = 1 - \frac{\sigma_M}{\sigma_u},$$

From Figure 1.27b, *draw the evolution of D as a function of the number of cycles N and compare in Figure 1.28 with the evolution of D calculated from the elasticity modulus change measured on Figure 1.27a*:

$$D = 1 - \frac{\tilde{E}}{E}.$$

Chapter 2

Thermodynamics and Micromechanics of Damage

A classical way to formulate in three dimensions the phenomena identified in one dimension as described for damage in Chapter 1 is to postulate the existence of energy potentials from which one can derive the state laws and the kinetic constitutive equations. In the thermodynamics of irreversible processes, two potentials are introduced and identified within the framework of the ''State Kinetic Coupling theory'' (J. Lemaitre, D. Marquis, 1988).

The state potential, written as a function of the state variables, defines the state laws and the variables associated with the state variables to define the power involved in each physical process. For the damage variable, an energy damage criterion is derived from the damaged elasticity potential.

The potential of dissipation written with the associated variables accounts for the kinetic laws of evolution of the flux dissipative variables. A constitutive equation for the damage gives the damage rate as a function of its associated variable.

This method is straightforward if the phenomena can be considered as isotropic. If not, the formulation at the mesoscale has to be guided by micromechanical analysis of basic mechanisms responsible for the anisotropy. The study of the growth of microcracks or microcavities restricts the degree of anisotropy to be introduced; the study of their behavior makes it possible to take crack closure effects into account and to analyze the degree of localization of the damage in many cases. It also justifies the choice of the continuous damage variable and accounts for a particular analytical expression of the kinetic law of damage evolution.

2.1 Three-Dimensional Analysis of Isotropic Damage

2.1.1 Thermodynamical Variables, State Potential

The method of local state in the thermodynamics of irreversible processes identifies one state variable per phenomenon to be analyzed. As the damage always occurs simultaneously with elastic and plastic strains, their classical variables must be taken into consideration.

Within the hypothesis of small strains and small displacements, the state variables at the mesoscale are divided as follows:

— Observable variables:
  * $\boldsymbol{\varepsilon}$ the total strain tensor of components ε_{ij} associated with the Cauchy stress tensor $\boldsymbol{\sigma}$ of components σ_{ij};
  * T the temperature associated with the entropy density s.

— Internal variables:

* $\boldsymbol{\varepsilon}^e$ the elastic strain tensor of component ε^e_{ij} associated with the stress tensor $\boldsymbol{\sigma}$;
* $\boldsymbol{\varepsilon}^p$ the plastic strain tensor of component ε^p_{ij}. It turns out that its associated variable is $-\boldsymbol{\sigma}$ due to the partitioning of strains:

$$\boldsymbol{\varepsilon} = \boldsymbol{\varepsilon}^e + \boldsymbol{\varepsilon}^p;$$

* r the damage accumulated plastic strain, whose associated variable is the isotropic strain hardening variable R. The von Mises criterion used in the plastic potential requires that its rate $\dot{r}$ be related to the damage and be proportional to the accumulated plastic strain rate defined by

$$\dot{p} = \left(\frac{2}{3}\dot{\varepsilon}^p_{ij}\dot{\varepsilon}^p_{ij}\right)^{1/2},$$

with the notation $\dot{p} = \dfrac{dp}{dt}$ and the summation convention of Einstein: $\dot{\varepsilon}^p_{ij}\dot{\varepsilon}^p_{ij} = \Sigma\left(\dot{\varepsilon}^p_{ij}\right)^2$
In Section 2.1.3 we shall also demonstrate that R measures the increase in radius of the yield cylinder in the stress space, this is an extension of the one-dimensional definition of R introduced in Section 1.2.4;
* $\boldsymbol{\alpha}$ or α_{ij} the back strain tensor, whose associated variable is the back stress $\mathbf{X}^D$ or X^D_{ij}. $\mathbf{X}^D$ represents the kinematic hardening, translation of the center of the yield surface in the deviatoric space, which makes that tensor a deviator: $tr(\mathbf{X}^D) = X^D_{kk} = 0$. Figure 2.1 gives the three-dimensional extension of Figure 1.12;
* D the damage variable. As defined in Section 1.2.1, $D = \delta S_D/\delta S$ depends a priori upon the plane $(\vec{n})$ of the section δS, but if the damage is considered

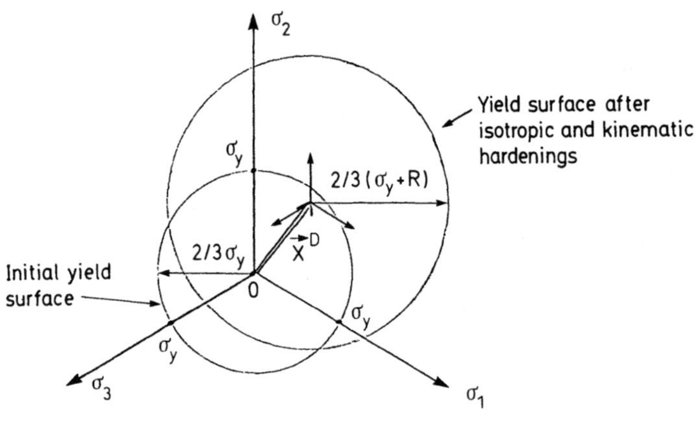

Fig. 2.1. Plastic yield locus in the principal stress space (The trisectrice is normal to the page!)

to be isotropic, it has the same value in all directions, and the scalar

$$D = \frac{\delta S_D}{\delta S} \quad \forall \vec{n}$$

characterizes completely the three-dimensional state of damage in the RVE at the point considered.
Let us call $\overline{Y}$ the associated variable of D which will be derived from the state potential. As D is dimensionless and since the product $-\overline{Y}\dot{D}$ is the power involved in the process of damage, we already see that $-\overline{Y}$ is a volume energy density.

Figure 2.2 is the chart of variables involved in the continuous mechanics of thermo-elasticity, plasticity and damage.
Let us postulate that the state laws are derived from a state potential: a continuous scalar function, concave with the temperature, convex with the other state variables and containing the origin. Taking the Helmholtz free energy,

$$\psi = \psi(\boldsymbol{\varepsilon}, T, \boldsymbol{\varepsilon}^e, \boldsymbol{\varepsilon}^p, r, \boldsymbol{\alpha}, D).$$

For elastoplasticity or elastoviscoplasticity, the strains act only through their difference $\boldsymbol{\varepsilon} - \boldsymbol{\varepsilon}^p = \boldsymbol{\varepsilon}^e$:

$$\psi = \psi([\boldsymbol{\varepsilon} - \boldsymbol{\varepsilon}^p], T, r, \boldsymbol{\alpha}, D) = \psi(\boldsymbol{\varepsilon}^e, T, r, \boldsymbol{\alpha}, D).$$

Assuming that the density ρ is constant, which is an approximation for ductile damage, the second principle of thermodynamics written as the Clausius-Duhem inequality gives the state laws (see Section 2.1.3):

$$\boldsymbol{\sigma} = \rho \frac{\partial \psi}{\partial \boldsymbol{\varepsilon}^e} \text{ and also } \boldsymbol{\sigma} = \rho \frac{\partial \psi}{\partial \boldsymbol{\varepsilon}} = -\rho \frac{\partial \psi}{\partial \boldsymbol{\varepsilon}^p};$$

$$s = -\frac{\partial \psi}{\partial T}.$$

State variables		Associated
observable	internal	variables
$\boldsymbol{\varepsilon}$		$\boldsymbol{\sigma}$
T		s
	$\boldsymbol{\varepsilon}^e$	$\boldsymbol{\sigma}$
	$\boldsymbol{\varepsilon}^p$	$-\boldsymbol{\sigma}$
	r	R
	$\boldsymbol{\alpha}$	$\mathbf{X}^D$
	D	$\overline{Y}$

Fig. 2.2. Chart of thermodynamic variables (J. Lemaitre 1976)

The associated variables are defined by

$$R = \rho\frac{\partial\psi}{\partial r},$$
$$\mathbf{X}^D = \rho\frac{\partial\psi}{\partial\boldsymbol{\alpha}},$$
$$\overline{Y} = \rho\frac{\partial\psi}{\partial D}$$

The analytical expression for ψ is chosen in keeping with the known experimental observations and the micromechanical results through the so called "State Kinetic Coupling theory":

— the linear isotropic elasticity;
— the state coupling of damage with elastic strain as shown experimentally in Section 1.2.4, and analytically by B. Budiansky and R. O'Connell (1976);
— the principle of strain equivalence stated in Section 1.2.3 of Chapter 1 together with the concept of effective stress written for the three-dimensional case as

$$\tilde{\boldsymbol{\sigma}} = \frac{\boldsymbol{\sigma}}{1-D} \text{ or } \tilde{\sigma}_{ij} = \frac{\sigma_{ij}}{1-D};$$

since the damage is isotropic, it has the same influence on all the components of the Cauchy stress tensor. For anisotropic damage, the variable is no longer a scalar and the effective stress needs the introduction of a more complex operator (see Section 2.2). The difference of the behavior in tension and compression is not taken into account here either, it will be, however, in Section 2.3.4.
— no state coupling neither between plasticity and elasticity nor between damage and plasticity,
— classical expressions for isotropic hardening and kinematic hardening giving saturation of hardening for large plastic strain (see Section 2.1.4),
— isothermal processes:

$$\psi = \frac{1}{\rho}\left[\frac{1}{2}a_{ijkl}\varepsilon^e_{ij}\varepsilon^e_{kl}(1-D) + R_\infty\left[r + \frac{1}{b}\exp(-br)\right] + \frac{X_\infty\gamma}{3}\alpha_{ij}\alpha_{ij}\right].$$

The law of elasticity coupled with damage is

$$\sigma_{ij} = \rho\frac{\partial\psi}{\partial\varepsilon_{ij}} = a_{ijkl}\varepsilon^e_{kl}(1-D),$$

where **a** is the fourth order elasticity stiffness tensor or, by inversion for the isotropic case,

$$\boxed{\varepsilon^e_{ij} = \frac{1+\nu}{E}\frac{\sigma_{ij}}{1-D} - \frac{\nu}{E}\frac{\sigma_{kk}}{1-D}\delta_{ij}},$$

E being Young's modulus, ν Poisson's ratio and δ_{ij} the Kronecker delta. Values for E and ν are given for some materials in Section 3.5.

The isotropic strain hardening scalar stress is expressed as:

$$R = \rho\frac{\partial\psi}{\partial r} = R_\infty[1-\exp(-br)].$$

R_∞ and b are two parameters which characterize the isotropic strain hardening for each material (see Section 5).

The kinematic strain hardening tensorial stress is expressed as

$$X_{ij}^D = \rho \frac{\partial \psi}{\partial \alpha_{ij}} = \frac{2}{3} X_\infty \gamma \alpha_{ij}.$$

X_∞ and γ are two parameters of nonlinear kinematic hardening written with the factor 2/3 to ensure a simple expression in one dimension. They must be identified for each material (see also Section 3.5).

The associated variable for D is defined as

$$\overline{Y} = \rho \frac{\partial \psi}{\partial D} = -\frac{1}{2} a_{ijkl} \varepsilon_{ij}^e \varepsilon_{kl}^e.$$

In order to work with a positive quantity, let us write

$$Y = -\overline{Y}.$$

Let us now find the relationship between Y and the elastic strain energy density w_e. By definition,

$$dw_e = \sigma_{ij} d\varepsilon_{ij}^e.$$

Integrating with the law of elasticity, and assuming no variation of damage, that is, D = const, yields.

$$w_e = \int a_{ijkl} \varepsilon_{kl}^e (1-D)\, d\varepsilon_{ij}^e = \frac{1}{2} a_{ijkl} \varepsilon_{ij}^e \varepsilon_{kl}^e (1-D).$$

This shows that

$$\boxed{Y = \frac{w_e}{1-D}}\,.$$

Y is also equal to one half the variation of the strain energy density corresponding to a variation of damage at constant stress: $d\sigma_{ij} = 0$ (J.L. Chaboche 1976).

Starting with the law of elasticity,

$$d\sigma_{ij} = a_{ijkl} \left[(1-D) d\varepsilon_{kl}^e - \varepsilon_{kl}^e dD \right] = 0,$$

or

$$d\varepsilon_{kl}^e = \varepsilon_{kl}^e \frac{dD}{1-D},$$

together with

$$dw_e|_{\sigma=\text{const}} = \sigma_{ij} d\varepsilon_{ij}^e = \sigma_{ij} \varepsilon_{ij}^e \frac{dD}{1-D} = a_{ijkl} \varepsilon_{ij}^e (1-D) \varepsilon_{kl}^e \frac{dD}{1-D}$$

or

$$\left.\frac{dw_e}{dD}\right|_{\sigma=\text{const}} = a_{ijkl} \varepsilon_{ij}^e \varepsilon_{kl}^e,$$

and from the definition of $Y = -\overline{Y}$

$$\boxed{Y = \frac{1}{2} \left.\frac{dw_e}{dD}\right|_{\sigma=\text{const}}}\,.$$

This last equation allows Y to be called *the strain energy density release rate*. This is the energy released by loss of stiffness of the RVE in which the damage occurred.

2.1.2 Damage Equivalent Stress Criterion

The strain energy density release rate, by definition of an associated variable in thermodynamics, is the principal variable which governs the phenomenon of damage. It is interesting to give it the meaning of an equivalent stress like the von Mises equivalent stress for plasticity.

We start with

$$Y = \frac{w_e}{1-D}.$$

Let us split w_e into two parts: the shear energy and the hydrostatic energy corresponding to the deviatoric part and the spherical part of the stress and strain tensors, respectively:

$$\sigma_{ij} = \sigma_{ij}^D + \sigma_H \delta_{ij},$$
$$\varepsilon_{ij}^e = \varepsilon_{ij}^{eD} + \varepsilon_H^e \delta_{ij}.$$

σ_H and ε_H^e are the hydrostatic stress and strain: $\sigma_H = \frac{1}{3}\sigma_{kk}$, $\varepsilon_H^e = \frac{1}{3}\varepsilon_{kk}^e$;

$$w_e = \int \sigma_{ij} d\varepsilon_{ij}^e = \int \sigma_{ij}^D d\varepsilon_{ij}^D + \delta_{ij}\delta_{ij} \int \sigma_H d\varepsilon_H^e.$$

From the law of linear isotropic elasticity coupled with damage,

$$\varepsilon_{ij}^{eD} = \frac{1+\nu}{E}\frac{\sigma_{ij}^D}{1-D}, \qquad \varepsilon_H^e = \frac{1-2\nu}{E}\frac{\sigma_H}{1-D}.$$

Remembering that $\delta_{ij}\delta_{ij} = 3(!)$

$$w_e = \frac{1}{2}\left(\frac{1+\nu}{E}\frac{\sigma_{ij}^D\sigma_{ij}^D}{1-D} + 3\frac{1-2\nu}{E}\frac{\sigma_H^2}{1-D}\right).$$

It is interesting to introduce the von Mises equivalent stress defined as

$$\sigma_{eq} = \left(\frac{3}{2}\sigma_{ij}^D\sigma_{ij}^D\right)^{1/2},$$

which leads Y to be equal to

$$Y = \frac{w_e}{1-D} = \frac{\sigma_{eq}^2}{2E(1-D)^2}\left[\frac{2}{3}(1+\nu) + 3(1-2\nu)\left(\frac{\sigma_H}{\sigma_{eq}}\right)^2\right].$$

$\frac{\partial_H}{\sigma_{eq}}$ is the triaxiality ratio, which plays a very important role in the rupture of materials, the measured ductility at fracture decreases as it increases. Remember what is known from practice: "High triaxiality makes materials brittle!"

Let us introduce

$$\boxed{R_\nu = \frac{2}{3}(1+\nu) + 3(1-2\nu)\left(\frac{\sigma_H}{\sigma_{eq}}\right)^2}$$

as the triaxiality function; then:

$$\boxed{Y = \frac{\tilde{\sigma}_{eq}^2}{2E} R_\nu}$$

The damage equivalent stress is now defined, similarly to the plastic equivalent stress, as the one-dimensional stress σ^* which, that for the same value of the damage, yields the same value of the elastic strain energy density as that of a three-dimensional state.

For the one-dimensional case:

$$[\sigma] = \begin{bmatrix} \sigma^* & 0 & 0 \\ 0 & 0 & 0 \\ 0 & 0 & 0 \end{bmatrix}, \quad \sigma_H = \frac{1}{3}\sigma^*, \quad [\sigma_D] = \begin{bmatrix} \frac{2}{3}\sigma^* & 0 & 0 \\ 0 & -\frac{1}{3}\sigma^* & 0 \\ 0 & 0 & -\frac{1}{3}\sigma^* \end{bmatrix},$$

$$\left.\begin{aligned} \sigma_{eq} &= \sigma^* \\ \frac{\sigma_H}{\sigma_{eq}} &= \frac{1}{3} \\ R_\nu &= 1 \end{aligned}\right\} w_e = \frac{\sigma^{*2}}{2E(1-D)}.$$

Writing the equality $w_e(\sigma^*) = w_e(\boldsymbol{\sigma}) = Y(1-D)$,

$$\frac{\sigma^{*2}}{2E(1-D)} = \frac{\sigma_{eq}^2 R_\nu}{2E(1-D)}$$

defines the damage equivalent stress as

$$\boxed{\sigma^* = \sigma_{eq} R_\nu^{1/2}}$$

This differs from the von Mises equivalent stress by the triaxiality function, which is in accordance with the physics described in Section 1.1: plasticity is mainly due to slips which do not depend upon the hydrostatic stress; damage is debonding influenced by the hydrostatic stress or the triaxiality ratio. This phenomenon is influenced by Poisson's ratio, which governs the elastic volume change; the factor R_ν increases in most cases when the Poisson's ratio decreases.

As an example, let us consider the damage equivalent stress for a case of plane stress:

$$[\sigma] = \begin{bmatrix} \sigma_1 & 0 & 0 \\ 0 & \sigma_2 & 0 \\ 0 & 0 & 0 \end{bmatrix}, \quad \sigma_H = \frac{1}{3}(\sigma_1 + \sigma_2),$$

$$[\sigma^D] = \frac{1}{3}\begin{bmatrix} 2\sigma_1 - \sigma_2 & 0 & 0 \\ 0 & 2\sigma_2 - \sigma_1 & 0 \\ 0 & 0 & -(\sigma_1 + \sigma_2) \end{bmatrix},$$

$$\sigma_{eq} = \left[\sigma_1^2 + \sigma_2^2 - \sigma_1\sigma_2\right]^{1/2},$$

$$\sigma^* = \left[\sigma_1^2 + \sigma_2^2 - \sigma_1\sigma_2\right]^{1/2}\left[\frac{2}{3}(1+\nu) + \frac{1}{3}(1-2\nu)\frac{(\sigma_1+\sigma_2)^2}{\sigma_1^2 + \sigma_2^2 - \sigma_1\sigma_2}\right]^{1/2}.$$

Figure 2.3 represents this expression in the plane $\left(\frac{\sigma_1}{\sigma^*}, \frac{\sigma_2}{\sigma^*}\right)$ for different values of Poisson's ratio.

2.1.3 Potential of Dissipation

Having defined all the state and associated variables, a second potential will give the kinetic constitutive relations to describe the evolution of the phenomena.
First of all, we must write the second principle of thermodynamics to ensure the validity of the model chosen. Let us start from the Clausius-Duhem inequality,

$$\sigma_{ij}\dot{\varepsilon}_{ij} - \rho(\dot{\psi} + s\dot{T}) - q_i \frac{T_{,i}}{T} \geq 0,$$

where $\vec{q}$ is the heat flux vector associated with the temperature gradient

$$\vec{\text{grad}}\ T \left(= \frac{\partial T}{\partial x_i} = T_{,i} \right)$$

for non isothermal processes.
The free energy is a function of all the state variables. Let us write its rate as

$$\dot{\psi} = \frac{\partial \psi}{\partial \varepsilon^e_{ij}} \dot{\varepsilon}^e_{ij} + \frac{\partial \psi}{\partial T}\dot{T} + \frac{\partial \psi}{\partial r}\dot{r} + \frac{\partial \psi}{\partial \alpha_{ij}}\dot{\alpha}_{ij} + \frac{\partial \psi}{\partial D}\dot{D}.$$

Together with $\dot{\varepsilon}_{ij} = \dot{\varepsilon}^e_{ij} + \dot{\varepsilon}^p_{ij}$ it becomes

$$\left(\sigma_{ij} - \rho\frac{\partial \psi}{\partial \varepsilon^e_{ij}}\right)\dot{\varepsilon}^e_{ij} - \rho\left(s + \frac{\partial \psi}{\partial T}\right)\dot{T} + \sigma_{ij}\dot{\varepsilon}^p_{ij} - \rho\frac{\partial \psi}{\partial r}\dot{r} - \rho\frac{\partial \psi}{\partial \alpha_{ij}}\dot{\alpha}_{ij}$$
$$-\rho\frac{\partial \psi}{\partial D}\dot{D} - q_i\frac{T_{,i}}{T} \geq 0$$

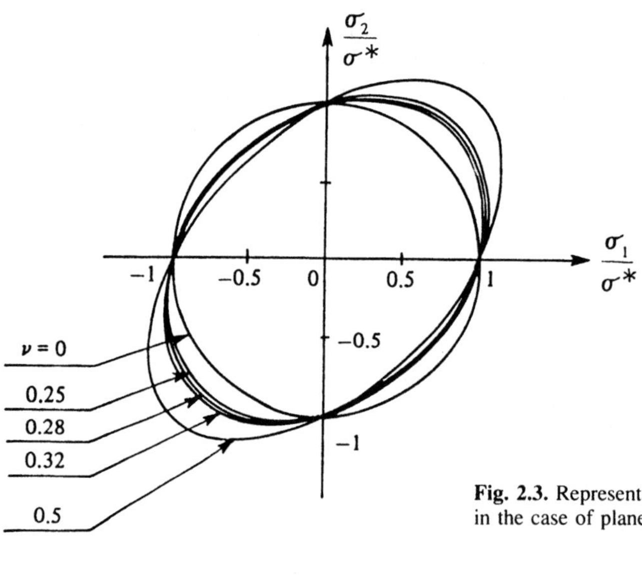

Fig. 2.3. Representation of the damage criterion in the case of plane stress (after D.Baptiste)

or, with the definition of the associated variables,

$$\sigma_{ij}\dot{\varepsilon}^p_{ij} - R\dot{r} - X^D_{ij}\dot{\alpha}_{ij} - \overline{Y}\dot{D} - q_i\frac{T_{,i}}{T} \geq 0.$$

To always satisfy this inequality of a positive dissipation and particularly for an isothermal process where the plastic dissipation is negligible, we must require that

$$-\overline{Y}\dot{D} \geq 0.$$

As $-\overline{Y}$ is a positive quadratic function, $-\overline{Y} \geq 0$, the damage rate $\dot{D}$ must be a non-negative function. The variable damage introduced can account only for deterioration of materials and not for recovery of strength.

Returning to the dissipation inequality, we see that it is the sum of the products of rate or flux variables (with the negative sign) multiplied by their dual variables as shown by the table in Figure 2.4.

Let us postulate that the kinetic laws are derived from a potential of dissipation, a scalar continuous and convex function F, of the dual variables, the state variables having the possibility to act as parameters.

For the isothermal case:

$$F(\underbrace{\boldsymbol{\sigma}, R, \mathbf{X}^D, Y}_{\text{variables}};\ \underbrace{(\boldsymbol{\varepsilon}^e, r, \boldsymbol{\alpha}, D)}_{\text{parameters}}).$$

The laws of plasticity or viscoplasticity coupled with damage are derived from this potential by means of a scalar multiplier which is always positive. This ensures the normality condition of yielding for plasticity, which does not depend upon time explicitly, and the viscosity property for the viscoplasticity. Within the so-called "normality rule of generalized standard materials,"

$$\left.\begin{aligned} \dot{\boldsymbol{\varepsilon}}^p &= \frac{\partial F}{\partial \boldsymbol{\sigma}}\dot{\lambda} \\ \dot{r} &= -\frac{\partial F}{\partial R}\dot{\lambda} \\ \dot{\boldsymbol{\alpha}} &= -\frac{\partial F}{\partial \mathbf{X}^D}\dot{\lambda} \end{aligned}\right\} \text{(Visco) Plasticity constitutive equations;}$$

Flux variables	Dual variables
$\dot{\boldsymbol{\varepsilon}}^p$	.... $\boldsymbol{\sigma}$
$-\dot{r}$	.... R
$-\dot{\boldsymbol{\alpha}}$	.... $\mathbf{X}^D$
$-\dot{D}$	.... $-\overline{Y}$
$\vec{q}$	...... $-\dfrac{\vec{\text{grad}}T}{T}$

Fig. 2.4. Table of dissipative variables

$$\dot{D} = -\frac{\partial F}{\partial \overline{Y}}\dot{\lambda} \rightarrow \text{ Kinetic law of damage evolution.}$$

In order to define the yield condition, we need a loading function which generalizes in three-dimensions the one-dimensional plasticity criterion in tension introduced in Section 1.2.4, namely,

$$\left|\frac{\sigma}{1-D} - X\right| - R - \sigma_y = 0$$

Let us work with the von Mises criterion which states that the (visco) plastic strains are governed by the elastic shear (or deviatoric) energy density, the so-called ''J_2 theory''. The corresponding equivalent stress is

$$\sigma_{eq} = \left(\frac{3}{2}\sigma_{ij}^D \sigma_{ij}^D\right)^{1/2},$$

$\boldsymbol{\sigma}^D$ being the stress deviator:$\sigma_{ij}^D = \sigma_{ij} - \frac{1}{3}\sigma_{kk}\delta_{ij}$. Together with kinematic hardening, the von Mises criterion is applied to define the size of the yield locus regardless of the translation $\mathbf{X}^D$ defined in Section 2.1.1 and shown in Figure 2.1. Therefore, it acts upon the difference between $\boldsymbol{\sigma}^D$ and $\mathbf{X}^D$.

Furthermore, in the presence of damage, the coupling between the damage and the plastic strain is written in accordance with the principle of strain equivalence (see Section 1.2.3). The yield criterion is written in the same way as for a nondamaged material except that the stress is replaced by the effective stress, which, for isotropic damage is:

$$\tilde{\boldsymbol{\sigma}} = \frac{\boldsymbol{\sigma}}{1-D}.$$

Then, the loading function f is written as

$$\boxed{f = \left(\tilde{\boldsymbol{\sigma}}^D - \mathbf{X}^D\right)_{eq} - R - \sigma_y}$$

with $$\left(\tilde{\boldsymbol{\sigma}}^D - \mathbf{X}^D\right)_{eq} = \left[\frac{3}{2}\left(\frac{\sigma_{ij}^D}{1-D} - X_{ij}^D\right)\left(\frac{\sigma_{ij}^D}{1-D} - X_{ij}^D\right)\right]^{1/2}.$$

Remember that $\mathbf{X}^D$ is a deviator and σ_y is the yield stress of each material which depends upon the temperature (see Section 3.5).

Plasticity (see Figure 2.5)

Plastic strain occurs only when the state of stress reaches the actual yield stress; this corresponds to the satisfaction of the yield criterion $f = 0$. The plastic strain continues to grow if the yield criterion is continuously satisfied, that is, if $\dot{f} = 0$. Then plasticity deals with these two conditions, which define loading, or unloading with $f \leq 0$:

$$\dot{\varepsilon}^p \neq 0 \text{ if } \begin{cases} f = 0 \\ \text{and} \\ \dot{f} = 0 \end{cases}, \quad \dot{\varepsilon}^p = 0 \text{ if } \begin{cases} f < 0 \\ \text{or} \\ \dot{f} < 0 \end{cases}.$$

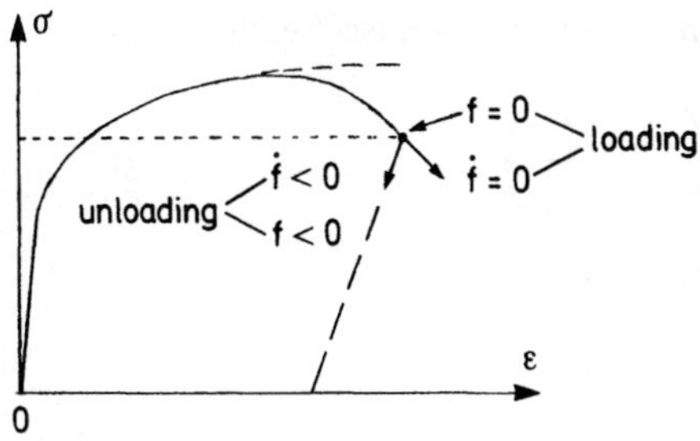

Fig. 2.5. Loading-unloading condition in one dimension for plasticity and damage

The expression of the multiplier $\dot{\lambda}$ is deduced from these two conditions $f = 0$ and $\dot{f} = 0$.

Viscoplasticity

The plastic strain becomes a time-dependent phenomenon as the temperature increases. For metals, this plastic viscosity is important if the loading occurs at a temperature above about a third of the melting temperature T_m:

$$T \geq \frac{1}{3} T_m \overset{\circ}{=} K.$$

For the uniaxial state of stress without any damage, Norton's law expresses the relation between the steady state plastic strain rate and the viscous stress:

$$\sigma_v = K_v \dot{\varepsilon}_p^{1/N},$$

where K_v and N are two material parameters. This relation is usually applied for low plastic strain rates and does not account for the saturation of viscosity where the strain rate is very high. Then for damage mechanics dealing with conditions close to failure, it is more appropriate to work with an expression like:

$$\sigma_v = K_\infty \left[1 - \exp\left(-\frac{\dot{\varepsilon}_p}{n} \right) \right] \text{ or } \dot{\varepsilon}_p = \mathrm{Ln}\left(1 - \frac{\sigma_v}{K_\infty} \right)^{-n},$$

where K_∞ and n are two material parameters (see Section 3.5)

$$(\dot{\varepsilon}^p \to \infty) \Rightarrow (\sigma_v \to K_\infty = \text{const}).$$

The stress $\tilde{\sigma}$ is now the sum of the yield stress, the strain hardening stresses R and X as for plasticity, and a viscous stress σ_v:

$$\tilde{\sigma} = \sigma_y + R + X + \sigma_v,$$

which means that the loading function f may be positive:

$$f = |\tilde{\sigma} - X| - R - \sigma_y = \sigma_v > 0.$$

Referring back to the multiplier $\dot{\lambda}$ of the three-dimensional case and using the von Mises function,

$$\dot{\varepsilon}_{ij}^p = \frac{\partial F}{\partial \sigma_{ij}} \dot{\lambda}, \quad f = \left[\left(\tilde{\boldsymbol{\sigma}}^D - \mathbf{X}^D \right)_{eq} - R - \sigma_y \right] > 0 \text{ for visco plasticity;}$$

by extension of the uniaxial case, $\dot{\lambda}$ is taken such that (see also Section 2.1.4)

$$\dot{p} = \frac{\dot{\lambda}}{1-D} = \mathrm{Ln}\left(1 - \frac{f}{K_\infty}\right)^{-n} .$$

Had using Norton's law, we would have obtained

$$\dot{p} = \left(\frac{f}{K_v}\right)^{N} .$$

2.1.4 Strain-Damage Coupled Constitutive Equations

An important problem in the mechanics of materials is the choice of an analytical expression for the two potentials and particularly for the potential of dissipation. Thermodynamics provides the general framework and some restrictions on the functions that can be used, but only experiments and micromechanics can give the details. As the constitutive equations must be general and valid for all materials (or almost!), we have to consider the general trends of basic experiments and micromechanics and let only the value of a few parameters to be characteristic of each material.
Let us look carefully at the results of a schematic experimental test in tension with some unloadings and compressive loadings, as shown in Figure 2.6.

1) As already mentioned in Section 1.2.4, below a certain value of the plastic strain, a threshold ε_{pD}, no damage occurs. This allows us to identify the (visco) plastic constitutive equations in their (large) domain of validity without considering any coupling with damage.
2) The kinematic back stress X, defined as the locus of the center of the elastic domain,
— increases with the plastic strain;

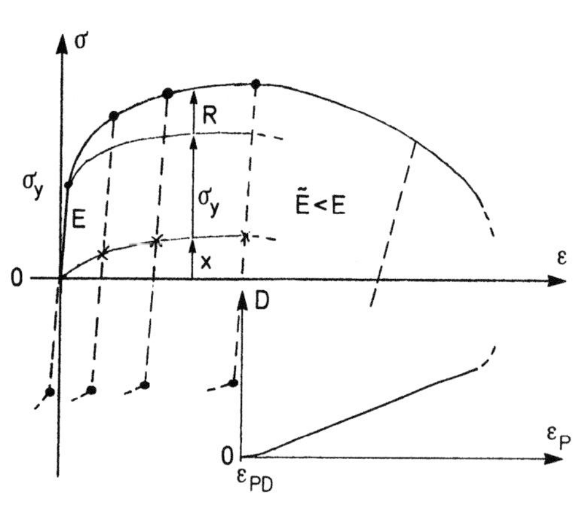

Fig. 2.6. General trends of kinematic and isotropic hardenings

— is nonlinear with the plastic strain;
— tends to saturate to some value X_∞.
This allows us to choose for the evolution of X an expression like

$$X = X_\infty \left[1 - \exp(-\gamma\varepsilon_p)\right],$$

which is more or less justified by physical and micromechanical considerations. X_∞ and γ are parameters to be identified for each material and each temperature (see table Section 3.5).

3) The isotropic hardening stress R, identified as

$$R = \sigma - \sigma_y - X$$

(where the yield stress σ_y is a constant),
— increases with the plastic strain;
— is nonlinear with the plastic strain;
— tends to also saturate to some value R_∞.
Let us take for the same (good!) reasons as for X,

$$R = R_\infty[1 - \exp(-b\varepsilon_p)],$$

as already written in Section 2.1.1 R_∞ and b are also parameters to be identified for each material and temperature.

Those equations, valid for one-dimensional tension, serve as a guide to writing the three-dimensional potential of dissipation where the kinetic coupling acts only by the effective stress deviator $\boldsymbol{\sigma}^D/(1-D)$. In the framework of the "State Kinetic Coupling theory", it corresponds to:

$$F = (\tilde{\boldsymbol{\sigma}}^D - \mathbf{X}^D)_{eq} - R - \boldsymbol{\sigma}_y + \frac{3}{4X_\infty} X_{ij}^D X_{ij}^D + F_{D(Y;(r,D))}.$$

The first three terms correspond to the loading function f. Working with a potential equal to the loading function $F = f$ gives rise to the so-called "associate plasticity theories." The fourth term $\frac{3}{4X_\infty} X_{ij}^D X_{ij}^D$ is responsible for the nonlinear kinematic hardening, including the factor 3/4 to ensure a simple expression in one dimension. The last term F_D is the part of the potential from which the kinetic law of damage evolution is derived. Experiments show that it does not depend explicitly upon $\boldsymbol{\sigma}$, $\mathbf{X}$ or R:

$$\dot{D} = -\frac{\partial F}{\partial \overline{Y}}\dot{\lambda} = \frac{\partial F_D}{\partial Y}\dot{\lambda}$$

The analytical expression for F_D will be discussed in detail in Chapter 3.
Now let us derive the general constitutive equations valid for plasticity and viscoplasticity:

$$\dot{\varepsilon}_{ij}^p = \frac{\partial F}{\partial \sigma_{ij}}\dot{\lambda} = \frac{\partial (\tilde{\boldsymbol{\sigma}}^D - \mathbf{X}^D)_{eq}}{\partial \sigma_{ij}}\dot{\lambda}$$

$$= \frac{\partial}{\partial \sigma_{ij}}\left(\frac{3}{2}\left[\frac{\sigma_{ij}^D}{1-D} - X_{ij}^D\right]\left[\frac{\sigma_{ij}^D}{1-D} - X_{ij}^D\right]\right)^{1/2}\dot{\lambda},$$

$$= \frac{1}{2} \frac{\frac{3}{2} \cdot 2 \left(\frac{\sigma_{ij}^D}{1-D} - X_{ij}^D \right) \frac{1}{1-D} - \frac{\partial X_{ij}^D}{\partial \sigma_{ij}}}{\left(\frac{\boldsymbol{\sigma}^D}{1-D} - \mathbf{X}^D \right)_{eq}} \dot{\lambda},$$

$$\dot{\varepsilon}_{ij}^p = \frac{3}{2} \frac{(\tilde{\sigma}_{ij}^D - X_{ij}^D)}{(\tilde{\boldsymbol{\sigma}}^D - \mathbf{X}^D)_{eq}} \frac{\dot{\lambda}}{1-D}.$$

An interesting variable in plasticity is the accumulated plastic strain p, defined by its rate $\dot{p}$:

$$\dot{p} = \left(\frac{2}{3} \dot{\varepsilon}_{ij}^p \dot{\varepsilon}_{ij}^p \right)^{1/2},$$

which is, for the one-dimensional case, simply the absolute value of the strain rate $\dot{\varepsilon}_p$:

$$\dot{p} = |\dot{\varepsilon}_p|.$$

Let us calculate $\dot{p}$ from the expression of $\dot{\varepsilon}_{ij}^p$:

$$\dot{p} = \left(\frac{2}{3} \dot{\varepsilon}_{ij}^p \dot{\varepsilon}_{ij}^p \right)^{1/2} = \sqrt{\frac{2}{3}} \cdot \frac{3}{2} \frac{\dot{\lambda}}{\left(\frac{\boldsymbol{\sigma}^D}{1-D} - \mathbf{X}^D \right)_{eq} (1-D)} \cdot$$

$$\cdot \left[\left(\frac{\sigma_{ij}^D}{1-D} - X_{ij}^D \right) \left(\frac{\sigma_{ij}^D}{1-D} - X_{ij}^D \right) \right]^{1/2},$$

which, together with the definition of

$$\left(\frac{\boldsymbol{\sigma}^D}{1-D} - \mathbf{X}^D \right)_{eq} = \left[\frac{3}{2} \left(\frac{\sigma_{ij}^D}{1-D} - X_{ij}^D \right) \left(\frac{\sigma_{ij}^D}{1-D} - X_{ij}^D \right) \right]^{1/2}$$

yields

$$\dot{p} = \frac{\dot{\lambda}}{1-D}.$$

The constitutive equation for the isotropic strain hardening is

$$\dot{r} = -\frac{\partial F}{\partial R} \dot{\lambda} = \dot{\lambda}.$$

This shows that

$$\boxed{\dot{r} = \dot{\lambda} = \dot{p}(1-D)}.$$

When no damage occurs, the state variable for isotropic hardening is the accumulated plastic strain,

$$D = 0 \rightarrow r = p.$$

For incremental calculations it is more convenient to work with a constitutive equation for the associated variable R. From the state potential,

$$R = R_\infty [1 - \exp(-br)]$$

or

$$\dot{R} = R_\infty b \exp(-br) \dot{r}$$

or

$$\dot{R} = b(R_\infty - R) \dot{\lambda}.$$

The last equation accounts for kinematic hardening:

$$\dot{\alpha}_{ij} = -\frac{\partial F}{\partial X^D_{ij}}\dot{\lambda} = -\left(\frac{\partial(\tilde{\boldsymbol{\sigma}}^D - \mathbf{X}^D)_{eq}}{\partial X^D_{ij}} + \frac{3}{2X_\infty}X^D_{ij}\right)\dot{\lambda}.$$

The same calculation as for $\dfrac{\partial F}{\partial \sigma_{ij}}$ gives

$$\dot{\alpha}_{ij} = -\left[\frac{3}{2}\frac{\left(\dfrac{\sigma^D_{ij}}{1-D} - X^D_{ij}\right)\dfrac{\partial(-X^D_{ij})}{\partial X^D_{ij}}}{\left(\dfrac{\boldsymbol{\sigma}^D}{1-D} - \mathbf{X}^D\right)_{eq}} + \frac{3}{2X_\infty}X^D_{ij}\right]\dot{\lambda},$$

$$\dot{\alpha}_{ij} = \dot{\varepsilon}^p_{ij}(1-D) - \frac{3}{2X_\infty}X^D_{ij}\dot{\lambda},$$

or, with the definition of $\mathbf{X}^D$ from the state potential in order to have a constitutive equation for this associated variable,

$$X^D_{ij} = \frac{2}{3}X_\infty\gamma\alpha_{ij} \quad \text{or} \quad \dot{X}^D_{ij} = \frac{2}{3}X_\infty\gamma\dot{\alpha}_{ij},$$

$$\dot{X}^D_{ij} = \gamma\left[\frac{2}{3}X_\infty\dot{\varepsilon}^p_{ij}(1-D) - X^D_{ij}\dot{\lambda}\right].$$

The last calculation to be performed is to find the multiplier $\dot{\lambda}$. (ouff!!)

Plasticity ($f = 0$)

The plastic multiplier is derived from the consistancy condition

$$\dot{f} = 0.$$

Let us start from

$$f = \left(\frac{\boldsymbol{\sigma}^D}{1-D} - \mathbf{X}^D\right)_{eq} - R - \sigma_y = 0,$$

$$\dot{f} = \frac{\partial f}{\partial \sigma_{ij}}\dot{\sigma}_{ij} + \frac{\partial f}{\partial X^D_{ij}}\dot{X}^D_{ij} + \frac{\partial f}{\partial D}\dot{D} + \frac{\partial f}{\partial R}\dot{R} = 0,$$

since σ_y is a material constant.

$\dfrac{\partial f}{\partial \sigma_{ij}}$ and $\dfrac{\partial f}{\partial X^D_{ij}}$ have already been calculated to determine $\dot{\varepsilon}^p_{ij}$ and $\dot{\alpha}_{ij}$:

$$\frac{\partial f}{\partial \sigma_{ij}} = \frac{\partial\left(\dfrac{\boldsymbol{\sigma}^D}{1-D} - \mathbf{X}^D\right)_{eq}}{\partial \sigma_{ij}} = \frac{3}{2}\frac{\left(\dfrac{\sigma^D_{ij}}{1-D} - X^D_{ij}\right)}{\left(\dfrac{\boldsymbol{\sigma}^D}{1-D} - \mathbf{X}^D\right)_{eq}}\frac{1}{1-D},$$

$$\frac{\partial f}{\partial X_{ij}^D} = \frac{\partial\left(\dfrac{\boldsymbol{\sigma}^D}{1-D} - \mathbf{X}^D\right)_{eq}}{\partial X_{ij}^D} = -\frac{3}{2}\frac{\left(\dfrac{\sigma_{ij}^D}{1-D} - X_{ij}^D\right)}{\left(\dfrac{\boldsymbol{\sigma}^D}{1-D} - \mathbf{X}^D\right)_{eq}},$$

$$\frac{\partial f}{\partial D} = \frac{\partial\left(\dfrac{\boldsymbol{\sigma}^D}{1-D} - \mathbf{X}^D\right)_{eq}}{\partial D} = \frac{\partial}{\partial D}\left(\frac{3}{2}\left[\frac{\sigma_{ij}^D}{1-D} - X_{ij}^D\right]\left[\frac{\sigma_{ij}^D}{1-D} - X_{ij}^D\right]\right)^{1/2},$$

$$= \frac{1}{2}\frac{\dfrac{3}{2}2\left(\dfrac{\sigma_{ij}^D}{1-D} - X_{ij}^D\right)\dfrac{\sigma_{ij}^D}{(1-D)^2}}{\left(\dfrac{\boldsymbol{\sigma}^D}{1-D} - \mathbf{X}^D\right)_{eq}},$$

$$\frac{\partial f}{\partial R} = -1.$$

Then:

$$\dot{f} = \frac{3}{2}\frac{\left(\dfrac{\sigma_{ij}^D}{1-D} - X_{ij}^D\right)}{\left(\dfrac{\boldsymbol{\sigma}^D}{1-D} - \mathbf{X}^D\right)_{eq}}\left(\frac{\dot{\sigma}_{ij}}{1-D} - \dot{X}_{ij}^D + \frac{\sigma_{ij}^D\dot{D}}{(1-D)^2}\right) - \dot{R} = 0.$$

Replacing $\dot{X}_{ij}^D$, $\dot{D}$ and $\dot{R}$ by

$$\dot{X}_{ij}^D = \frac{2}{3}X_\infty\gamma\dot{\varepsilon}_{ij}^p(1-D) - \gamma X_{ij}^D\dot{\lambda} = X_\infty\gamma\frac{\left(\dfrac{\sigma_{ij}^D}{1-D} - X_{ij}^D\right)}{\left(\dfrac{\boldsymbol{\sigma}^D}{1-D} - \mathbf{X}^D\right)_{eq}}\dot{\lambda} - \gamma X_{ij}^D\dot{\lambda},$$

$$\dot{D} = -\frac{\partial F_D}{\partial \overline{Y}}\dot{\lambda} = \frac{\partial F_D}{\partial Y}\dot{\lambda};$$

$$\dot{R} = b(R_\infty - R)\dot{\lambda}.$$

This allows us to calculate $\dot{\lambda}$:

$$\dot{f} = \frac{3}{2}\frac{\left(\dfrac{\sigma_{ij}^D}{1-D} - X_{ij}^D\right)\dot{\sigma}_{ij}}{\left(\dfrac{\boldsymbol{\sigma}^D}{1-D} - \mathbf{X}^D\right)_{eq}(1-D)} -$$

$$-\left[X_\infty\gamma + \frac{3}{2}\frac{\left(\dfrac{\sigma_{ij}^D}{1-D} - \mathbf{X}_{ij}^D\right)}{\left(\dfrac{\boldsymbol{\sigma}^D}{1-D} - \mathbf{X}^D\right)_{eq}}\left(-\gamma X_{ij}^D - \frac{\sigma_{ij}^D}{(1-D)^2}\frac{\partial F_D}{\partial Y}\right) + b(R_\infty - R)\right]\dot{\lambda}=0$$

$$\dot{\lambda} = \frac{3}{2}\left(\frac{\sigma_{ij}^D}{1-D} - X_{ij}^D\right)\dot{\sigma}_{ij} \Bigg/ \Bigg\{ \left(\frac{\boldsymbol{\sigma}^D}{1-D} - \mathbf{X}^D\right)_{eq} (1-D) \cdot$$

$$\cdot \left[X_\infty \gamma - \frac{3}{2} \frac{\left(\frac{\sigma_{ij}^D}{1-D} - X_{ij}^D\right)}{\left(\frac{\boldsymbol{\sigma}^D}{1-D} - \mathbf{X}^D\right)_{eq}} \left(\gamma X_{ij}^D + \frac{\sigma_{ij}^D}{(1-D)^2}\frac{\partial F_D}{\partial Y}\right) + b(R_\infty - R) \right] \Bigg\},$$

or, in a more compact form with $\tilde{\boldsymbol{\sigma}} = \dfrac{\boldsymbol{\sigma}}{1-D}$

$$\dot{\lambda} = \frac{3}{2}\left(\tilde{\sigma}_{ij}^D - X_{ij}^D\right)\dot{\sigma}_{ij} \Bigg/ \Bigg\{ \left(\tilde{\boldsymbol{\sigma}}^D - \mathbf{X}^D\right)_{eq} (1-D) \cdot$$

$$\cdot \left[X_\infty \gamma - \frac{3}{2} \frac{\left(\tilde{\sigma}_{ij}^D - X_{ij}^D\right)}{\left(\tilde{\boldsymbol{\sigma}}^D - \mathbf{X}^D\right)_{eq}} \left(\gamma X_{ij}^D + \frac{\tilde{\sigma}_{ij}^D}{1-D}\frac{\partial F_D}{\partial Y}\right) + b(R_\infty - R) \right] \Bigg\},$$

The damage influences the plasticity by the effective deviatoric stress $\tilde{\sigma}_{ij}^D = \dfrac{\sigma_{ij}^D}{1-D}$ and by the partial derivative of the damage potential $\dfrac{\partial F_D}{\partial Y}$.

Viscoplasticity $(f > 0)$ (A. Benallal 1985)

This case is much simpler, as the loading function gives the viscosity effect by the viscosity law:

$$\dot{p} = \frac{\dot{\lambda}}{1-D} = \mathrm{Ln}\left(1 - \frac{f}{K_\infty}\right)^{-n},$$

$$\frac{\dot{\lambda}}{1-D} = \mathrm{Ln}\left[1 - \frac{\left(\tilde{\boldsymbol{\sigma}}^D - \mathbf{X}^D\right)_{eq} - R - \sigma_y}{K_\infty}\right]^{-n}.$$

Let us summarize the complete set of constitutive equations of the elasto-(visco)-plasticity coupled with damage: (The case of perfectly plastic materials is treated in Section 3.1.4)

$$\boxed{\begin{aligned}
\varepsilon_{ij} &= \varepsilon^e_{ij} + \varepsilon^p_{ij},\\
\varepsilon^e_{ij} &= \frac{1+\nu}{E}\tilde{\sigma}_{ij} - \frac{\nu}{E}\tilde{\sigma}_{kk}\delta_{ij},\\
\dot{\varepsilon}^p_{ij} &= \frac{3}{2}\frac{\tilde{\sigma}^D_{ij} - X^D_{ij}}{(\tilde{\boldsymbol{\sigma}}^D - \mathbf{X}^D)_{eq}}\frac{\dot{\lambda}}{1-D},\\
\dot{R} &= b(R_\infty - R)\dot{\lambda},\\
\dot{X}^D_{ij} &= \gamma\left[\frac{2}{3}X_\infty\dot{\varepsilon}^p_{ij}(1-D) - X^D_{ij}\dot{\lambda}\right],\\
\dot{D} &= \frac{\partial F_D}{\partial Y}\dot{\lambda},
\end{aligned}}$$

with

$$\tilde{\sigma}_{ij} = \frac{\sigma_{ij}}{1-D},$$

$$(\tilde{\boldsymbol{\sigma}} - \mathbf{X}^D)_{eq} = \left[\frac{3}{2}\left(\frac{\sigma^D_{ij}}{1-D} - X^D_{ij}\right)\left(\frac{\sigma^D_{ij}}{1-D} - X^D_{ij}\right)\right]^{1/2},$$

$$\sigma^D_{ij} = \sigma_{ij} - \frac{1}{3}\sigma_{kk}\delta_{ij},$$

$$f = (\tilde{\boldsymbol{\sigma}}^D - \mathbf{X}^D)_{eq} - R - \sigma_y.$$

Plasticity $(f \leq 0)$

if $f = 0$ and $\dot{f} = 0$

$$\dot{\lambda} = \frac{3}{2}\left(\tilde{\sigma}^D_{ij} - X^D_{ij}\right)\dot{\sigma}_{ij} \Big/ (1-D)\left[(\tilde{\boldsymbol{\sigma}}^D - \mathbf{X}^D)_{eq}[X_\infty\gamma + b(R_\infty - R)] - \frac{3}{2}(\tilde{\sigma}^D_{ij} - X^D_{ij})\left(\frac{\tilde{\sigma}^D_{ij}}{1-D}\frac{\partial F_D}{\partial Y} + \gamma X^D_{ij}\right)\right]$$

if $f < 0$ or $\dot{f} = 0$

$$\dot{\lambda} = 0$$

Viscoplasticity $(f > 0)$

$$\boxed{\frac{\dot{\lambda}}{1-D} = \mathrm{Ln}\left[1 - \frac{(\tilde{\boldsymbol{\sigma}}^D - \mathbf{X}^D)_{eq} - R - \sigma_y}{K_\infty}\right]^{-n}}$$

Identification of the material parameters

All the characteristic coefficients of the materials may be identified from one-dimensional experiments in tension compression. Let us write this interesting case in which damage does not exist, $D = 0$:

$$[\sigma] = \sigma \begin{bmatrix} 1 & 0 & 0 \\ 0 & 0 & 0 \\ 0 & 0 & 0 \end{bmatrix} \quad [\varepsilon] = \varepsilon_e \begin{bmatrix} 1 & 0 & 0 \\ 0 & -\nu & 0 \\ 0 & 0 & -\nu \end{bmatrix} + \varepsilon_p \begin{bmatrix} 1 & 0 & 0 \\ 0 & -\frac{1}{2} & 0 \\ 0 & 0 & -\frac{1}{2} \end{bmatrix},$$

$$[\sigma^D] = \sigma \begin{bmatrix} \frac{2}{3} & 0 & 0 \\ 0 & -\frac{1}{3} & 0 \\ 0 & 0 & -\frac{1}{3} \end{bmatrix} \quad [X^D] = X \begin{bmatrix} \frac{2}{3} & 0 & 0 \\ 0 & -\frac{1}{3} & 0 \\ 0 & 0 & -\frac{1}{3} \end{bmatrix}.$$

From the constitutive equations obtained previously with $D = 0$ and $\dot{D} = 0$,

$$(\boldsymbol{\sigma}^D - \mathbf{X}^D)_{eq} = \left[\frac{3}{2}\left(\frac{4}{9}(\sigma - X)^2 + \frac{1}{9}(\sigma - X)^2 + \frac{1}{9}(\sigma - X)^2\right)\right]^{1/2} = |\sigma - X|,$$

$$f = |\sigma - X| - R - \sigma_y,$$

$$\varepsilon = \varepsilon_e + \varepsilon_p,$$

$$\varepsilon_e = \frac{\sigma}{E},$$

$$\dot{\varepsilon}_p = \frac{\sigma - X}{|\sigma - X|}\dot{\lambda} = \dot{\lambda}\,\mathrm{sign}(\sigma - X),$$

$$\dot{R} = b(R_\infty - R)\dot{\lambda},$$

$$\frac{2}{3}\dot{X} = \frac{2}{3}X_\infty \gamma \dot{\varepsilon}_p - \frac{2}{3}\gamma X \dot{\lambda},$$

$$\dot{\lambda} = \dot{r} = \dot{p} = |\dot{\varepsilon}_p|.$$

For plasticity, $$\dot{\lambda} = \frac{\frac{3}{2}\frac{2}{3}(\sigma - X)\dot{\sigma}}{|\sigma - X|\,[X_\infty \gamma + b(R_\infty - R)] - \frac{3}{2}\frac{2}{3}(\sigma - X)\gamma X},$$

or $$\dot{\lambda} = \frac{\dot{\sigma}}{[X_\infty \gamma + b(R_\infty - R)]\,\mathrm{sign}(\sigma - X) - \gamma X}.$$

For viscoplasticity, $$\dot{\lambda} = \mathrm{Ln}\left[1 - \frac{|\sigma - X| - R - \sigma_y}{K_\infty}\right]^{-n}.$$

The final equations are:

$$\varepsilon = \varepsilon_e + \varepsilon_p,$$

$$\varepsilon_e = \frac{\sigma}{E},$$

$$\dot{R} = b(R_\infty - R)\,|\dot{\varepsilon}_p|,$$

$$\dot{X} = \gamma\left(X_{\infty}\dot{\varepsilon}_p - X\left|\dot{\varepsilon}_p\right|\right);$$

$$\dot{\varepsilon}_p = \frac{\dot{\sigma}}{X_{\infty}\gamma + b(R_{\infty} - R) - \gamma X \operatorname{sign}(\sigma - X)} \quad \text{for plasticity}$$

$$\dot{\varepsilon}_p = \mathrm{Ln}\left[1 - \frac{|\sigma - X| - R - \sigma_y}{K_{\infty}}\right]^{-n} \quad \text{for viscoplasticity}$$

For monotonic loading ($\left|\dot{\varepsilon}_p\right| = \dot{\varepsilon}_p$) the two strain-hardening equations may be integrated with the initial condition $\varepsilon_p = 0 \rightarrow R = 0$ and $X = 0$:

$$\frac{dR}{d\varepsilon_p} + bR = bR_{\infty} \rightarrow R = R_{\infty}[1 - \exp(-b\varepsilon_p)],$$

$$\frac{dX}{d\varepsilon_p} + \gamma X = \gamma X_{\infty} \rightarrow X = X_{\infty}[1 - \exp(-\gamma\varepsilon_p)].$$

The complete set of parameters to be identified are:

— E and ν for elasticity by classical methods;

— σ_y the yield stress taken as the stress which produces the first offset of plastic strain defined on a tensile test as;

$$\frac{\delta\varepsilon_p}{\varepsilon_e} \simeq 10\%;$$

— X_{∞} and γ for the kinematic hardening. The partition of stresses is made as shown in Figure 2.6, the plot of X as a function of ε_p allows us to identify X_{∞} and γ either on a semi-Log diagram or by numerical identification through the least square method, for example;

— R and b for the isotropic hardening by the same method.
A schematic example is given in Figure 2.7.

For viscoplasticity two more parameters need to be identified. The most convenient method is to derive them from a relaxation test at constant strain for a large value of the plastic strain when the strain hardening is almost saturated ($X \simeq X_{\infty}, R \simeq R_{\infty}$) and when the damage has not yet occurred:

$$\varepsilon = \text{const} \rightarrow \dot{\varepsilon} = 0 \rightarrow \dot{\varepsilon}_e + \dot{\varepsilon}_p = 0 \rightarrow \dot{\varepsilon}_p = -\frac{\dot{\sigma}}{E};$$

for $\sigma > X_{\infty} + R_{\infty} + \sigma_y$,

$$\dot{\varepsilon}_p = -\frac{\dot{\sigma}}{E} = \mathrm{Ln}\left(1 - \frac{\sigma - X_{\infty} - R_{\infty} - \sigma_y}{K_{\infty}}\right)^{-n}.$$

If the stress is recorded as a function of the time and if $(X_{\infty} + R_{\infty} + \sigma_y)$ is known from the above method for a very low strain rate tension compression test, one may calculate $\dot{\varepsilon}_p$ and plot it as a function of σ; then n and K_{∞} are the values which

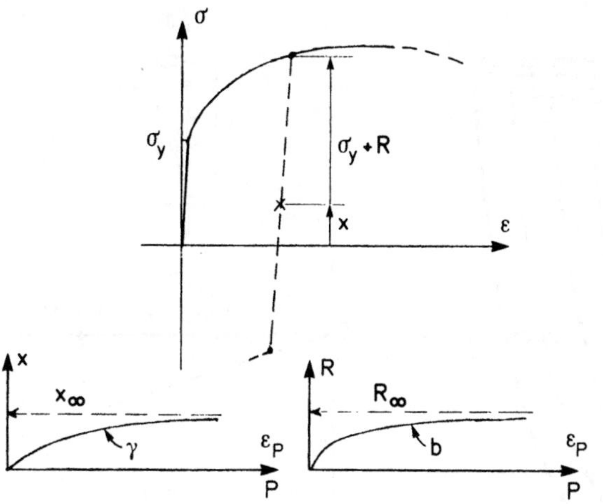

Fig. 2.7. Identification of elastoplastic parameters

fit best with the experimental points. A systematic method consists in a numerical identification of both coefficients n and K_∞ using the least square method.

2.1.5 Exercise on the Identification of Material Parameters

A result of an experimental test in tension compression is given in Figure 2.8. It gives the stabilized stress-strain loops after 5 cycles for different amplitudes of strain, and also the radial strain to determine Poisson's ratio.
Please follow carefully the method described in Section 2.1.4 under "Identification of the material parameters" in order *to derive the values of the elastoplastic parameters*, you will be glad to find the following answers!

Elasticity $E \simeq 200\,000$ MPa
$\nu \simeq 0.3$

Plasticity $\sigma_y \simeq 260$ MPa
$X_\infty \simeq 200$ MPa
$\gamma \simeq 2$
$R_\infty \simeq 300$ MPa
$b \simeq 1$

Check that the relative error on predictions remains within 10% on stresses

2.2 Analysis of Anisotropic Damage

Assuming the damage to be isotropic is realistic in many cases, especially under conditions of proportional loading when the principal directions of the stresses remain constant. This assumption allows us to predict the conditions for crack initiation with an accuracy often admissible in engineering applications where a prediction of the lifetime with an uncertainty of 10 to 50% is already very helpful.

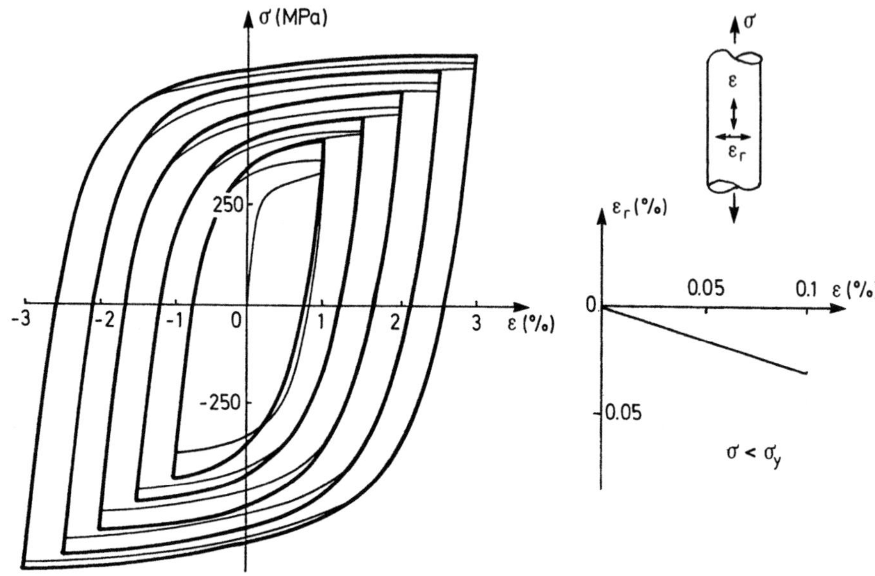

Fig. 2.8. Cyclic test result on AISI 316 L stainless steel at room temperature (after J.L. Chaboche)

Nevertheless all materials under special loading conditions and most brittle materials develop anisotropic damage for which the damage variable can no longer be a scalar.

Maximum principal stress damage concept (F. Leckie, D. Hayhurst 1973)

A simple way to model anisotropic damage is to assume that it occurs only in the plane normal to the maximum principal stress. The damage is then characterized by its intensity D and by the unit vector of the orientation of the maximum principal stress $\vec{n}_p$. It is a vector: $\vec{D} = D\vec{n}_p$.
The effective stress is written in the principal stress coordinates:

$$[\tilde{\sigma}] = \begin{bmatrix} \sigma_1 & 0 & 0 \\ 0 & \sigma_2 & 0 \\ 0 & 0 & \dfrac{\sigma_3}{1-D} \end{bmatrix} \quad \text{if } \sigma_1 < \sigma_2 < \sigma_3.$$

2.2.1 Geometrical Definition of a Second-Order Damage Tensor (S. Murakami 1981)

For a more complex type of anisotropy and for use in nonproportional loading, it is possible to extend the surface definition of the damage introduced for one dimension in Section 1.2.1
Recall that if δS represents the surface area of the current configuration with damage of surface δS_D, then $\delta S - \delta S_D = \delta S(1 - D)$ represents the nondamaged

area of the strain-equivalent undamaged configuration. Let us develop this idea for anisotropic damage.

Consider a RVE of a damaged material unloaded in its current configuration and a sectional area defined by its orientation $\vec{n}$, its surface δS and its shape (Figure 2.9). Consider also the strain-equivalent undamaged configuration in the same coordinate system. The R.V.E. has changed and particularly the sectional area has been transformed into a section of different orientation $\tilde{\vec{n}}$ due to the anisotropy of damage and of the smaller area $\delta\tilde{S}$ (Figure 2.9). Let us assume that the shape remains the same; later on we shall see that this hypothesis corresponds to orthotropic damage.

$$\vec{n}\delta S \text{ is transformed into } \tilde{\vec{n}}\delta\tilde{S}$$

By analogy with the one-dimensional case, the second-order tensor which transforms the vector $\vec{n}\delta S$ into the vector $\tilde{\vec{n}}\delta\tilde{S}$ characterizes the damage and is denoted by $(\mathbf{I}-\mathbf{D})$ where $\mathbf{I}$ is the unit second order tensor

$$(\mathbf{I}-\mathbf{D})\cdot\vec{n}\delta S = \tilde{\vec{n}}\delta\tilde{S} \quad \text{or} \quad (I_{ij}-D_{ij})n_j\delta S = \tilde{n}_i\delta\tilde{S}$$

D is the second-order tensor damage variable. It is symmetric and has the three principal values D_1, D_2, D_3 in the principal coordinates (X_1, X_2, X_3) which corresponds to the reduction of the areas of a tetrahedron drawn in the X_1, X_2, X_3

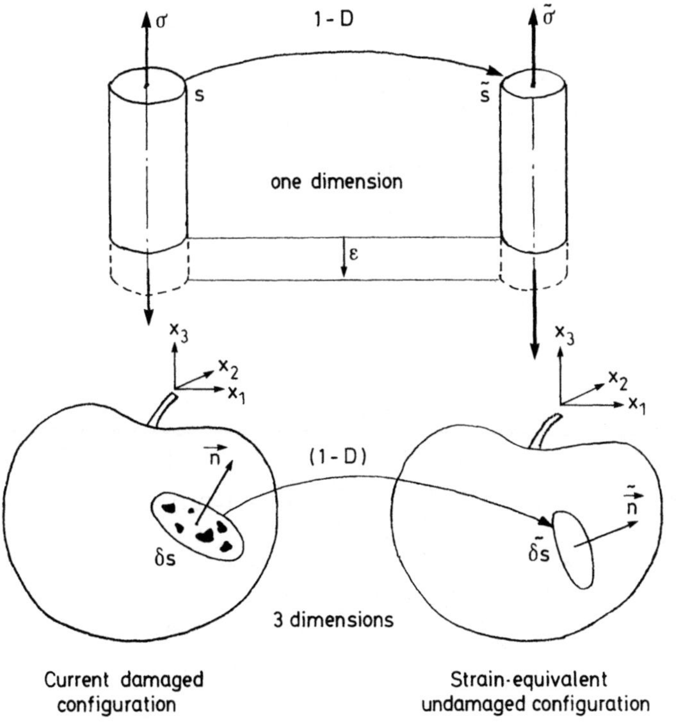

Fig. 2.9. Isotropic damage in one dimension and anisotropic damage in three dimensions.

coordinates. Let PQR be a triangle of area δS in a tetrahedron $OPQR$ in the current configuration which transforms into a triangle $\tilde{P}\tilde{Q}\tilde{R}$ of area $\delta\tilde{S}$ in the strain-equivalent undamaged configuration (Figure 2.10) then we have:

$$\begin{aligned} & PQR \to \tilde{P}\tilde{Q}\tilde{R} \to (I_{ij} - D_{ij})\cdot n_j \delta S = \tilde{n}_i \delta\tilde{S}; \\ \vec{n} = \vec{X}_1 \quad & OQR \to O\tilde{Q}\tilde{R} \to (1 - D_1)\delta A_1 = \delta\tilde{A}_1, \\ \vec{n} = \vec{X}_2 \quad & ORP \to O\tilde{R}\tilde{P} \to (1 - D_2)\delta A_2 = \delta\tilde{A}_2, \\ \vec{n} = \vec{X}_3 \quad & OPQ \to O\tilde{P}\tilde{Q} \to (1 - D_3)\delta A_3 = \delta\tilde{A}_3. \end{aligned}$$

This property of symmetry with respect to three planes makes the damage orthotropic. Nevertheless it covers a wide range of applications.
To define the effective stress, let us first write the stress vector $\vec{T}$ in the two configurations. For the one-dimensional case;

$$T_i \delta S = \tilde{T}_i \delta\tilde{S}.$$

Introducing the effective stress tensor $\tilde{\sigma}$ as

$$\tilde{\sigma}_{ij}\tilde{n}_j = \tilde{T}_i,$$

we can write

$$\sigma_{ij} n_j \delta S = \tilde{\sigma}_{ij}\tilde{n}_j \delta\tilde{S},$$

or with the definition of $(\mathbf{I} - \mathbf{D})$,

$$\begin{aligned} \tilde{n}_j \delta\tilde{S} &= (I_{ij} - D_{ij}) n_i \delta S, \\ \sigma_{ij} n_j &= \tilde{\sigma}_{ij}(I_{kj} - D_{kj}) n_k \quad \forall \vec{n}. \end{aligned}$$

Then, the effective stress tensor could be defined by

$$\tilde{\boldsymbol{\sigma}} = \boldsymbol{\sigma} \cdot (\mathbf{I} - \mathbf{D})^{-1},$$

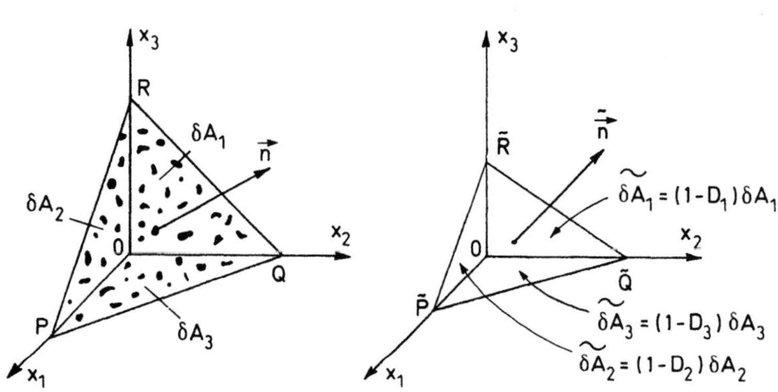

Fig. 2.10. Principal values of the damage tensor

but this leads to a nonsymmetric tensor. As only the symmetric part accounts for the constitutive equations of elasticity and plasticity, let us consider

$$\tilde{\boldsymbol{\sigma}} = \frac{1}{2}\left[\boldsymbol{\sigma}\cdot(\mathbf{I}-\mathbf{D})^{-1}+(\mathbf{I}-\mathbf{D})^{-1}\cdot\boldsymbol{\sigma}\right].$$

When the principal directions of stress and damage coincide (see Exercise 2.2.4.),

$$[\tilde{\sigma}] = \begin{bmatrix} \dfrac{\sigma_1}{1-D_1} & & \\ 0 & \dfrac{\sigma_2}{1-D_2} & 0 \\ 0 & 0 & \dfrac{\sigma_3}{1-D_3} \end{bmatrix}.$$

The definition of **D** as a symmetric second order tensor and of its associated effective stress are the basis for developing a complete theory of the orthotropic damage mechanics in the same maner as for the isotropic case. It is only "a bit" more complicated! In particular, the associate variable also becomes a second-order tensor **Y**.

The Helmholtz free energy state potential for linear elasticity coupled with damage is written as a scalar in a form similar to that in Section 2.1.1:

$$\psi = \frac{1}{\rho}\left[\frac{1}{2}(I_{ij}-D_{ij})a_{imkl}\varepsilon^e_{jm}\varepsilon^e_{kl}\right]+\psi_p,$$

where ψ_p is the plastic part which does not depend upon **D**.

Then

$$\overline{Y}_{ij} = \rho\frac{\partial\psi}{\partial D_{ij}} = -\frac{1}{2}a_{imkl}\varepsilon^e_{jm}\varepsilon^e_{kl},$$

or

$$Y_{ij} = \frac{1}{2}a_{imkl}\varepsilon^e_{jm}\varepsilon^e_{kl}.$$

2.2.2 Thermodynamical Definition of a Fourth-Order Damage Tensor (J.L. Chaboche 1978)

Another way to define the damage variable is to use the concept of effective stress associated with the principle of strain equivalence for linear elasticity.

Remember that for the one-dimensional case the law of elasticity coupled to damage is

$$\tilde{\sigma} = E\varepsilon_e,$$

or

$$\sigma = \tilde{E}\varepsilon_e,$$

for which

$$(1-D) = \frac{\tilde{E}}{E}.$$

Following the same ideas, the three-dimensional law of elasticity of a damaged RVE may be written, using the principle of strain equivalence, as

$$\tilde{\sigma}_{ij} = a_{ijkl}\varepsilon^e_{kl}.$$

It may also be derived from the state potential:

$$\psi = \frac{1}{2\rho}\tilde{a}_{ijkl}\varepsilon^e_{ij}\varepsilon^e_{kl} + \psi_p,$$

$$\sigma_{ij} = \rho\frac{\partial\psi}{\partial\varepsilon^e_{ij}} = \tilde{a}_{ijkl}\varepsilon^e_{kl},$$

where $\tilde{\mathbf{a}}$ is the elasticity tensor modified by damage. Assuming it has an inverse,

$$\varepsilon^e_{kl} = \tilde{a}^{-1}_{klij}\sigma_{ij}.$$

Substituting in the previous equation for $\tilde{\sigma}_{ij}$ with a change of dummy indices,

$$\tilde{\sigma}_{ij} = a_{ijkl}\tilde{a}^{-1}_{klrs}\sigma_{rs}.$$

By definition of the effective stress, $(a_{ijkl}\cdot\tilde{a}^{-1}_{klrs})$ is the operator of damage. It is a fourth-order tensor, by analogy with the one-dimensional case let us call it:

$$(I_{ijkl} - D_{ijkl})^{-1} = a_{ijrs}\tilde{a}^{-1}_{rskl},$$

where D_{ijkl} are the components of the fourth order tensor damage variable $\mathbf{D}$, and I_{ijkl} are the components of the fourth-order unit tensor $\mathbf{I}$ for symmetrical fourth order tensors:

$$D_{ijkl} = I_{ijkl} - \tilde{a}_{ijrs}a^{-1}_{rskl},$$

or

$$\mathbf{D} = \mathbf{I} - \tilde{\boldsymbol{a}} : \boldsymbol{a}^{-1},$$

where the symbol (:) means the contracted product upon two indices.

The associated variable $\mathbf{Y}$ is also a fourth order tensor:

$$\psi = \frac{1}{2\rho}\tilde{a}_{ijkl}\varepsilon^e_{ij}\varepsilon^e_{kl} + \psi_p,$$

with

$$\tilde{a}_{ijkl} = (I_{ijrs} - D_{ijrs})a_{rskl} = a_{ijkl} - D_{ijrs}a_{rskl};$$

$$\psi = \frac{1}{2\rho}(I_{ijrs} - D_{ijrs})a_{rskl}\varepsilon^e_{ij}\varepsilon^e_{kl} + \psi_p,$$

or

$$\overline{Y}_{ijkl} = \rho\frac{\partial\psi}{\partial D_{ijkl}} = -\frac{1}{2}a_{klrs}\varepsilon^e_{ij}\varepsilon^e_{rs},$$

$$Y_{ijkl} = \frac{1}{2}a_{klrs}\varepsilon^e_{ij}\varepsilon^e_{rs}.$$

This theory corresponds to the general case of anisotropy. We shall see in Section 2.3.3 that micromechanics of a general pattern of microcracks leads to damage representation as a combination of a scalar, a second-order tensor and a fourth-order tensor.

2.2.3 Energetic Definition of a Double Scalar Variable (P. Ladeveze, 1983)

For materials such as composites, in which the mechanisms of damage may be different in tension and shear it is interesting to develop a theory defining two independent scalar variables which characterize the damage that influences the elastic shear energy and the elastic hydrostatic energy:

— D_s acts on the deviatoric stress components;
— D_n acts on the hydrostatic stress.

The damage may be isotropic or anisotropic if the material is not subjected to damage in certain material directions.
For the general case of anisotropic elasticity, but without any damage, the complementary energy density is

$$w_e^* = \frac{1}{2} S_{ijkl} \sigma_{ij} \sigma_{kl},$$

where $\mathbf{S}$ is the compliance fourth-order tensor $\mathbf{S} = \mathbf{a}^{-1}$. It may be split into two parts as in Section 2.1.2 by means of the stress deviator $\boldsymbol{\sigma}_D$ and the hydrostatic stress σ_H:

$$\sigma_H = \frac{1}{3} \sigma_{kk},$$

$$\sigma_{ij}^D = \sigma_{ij} - \sigma_H \delta_{ij},$$

$$w_e^* = \underbrace{\frac{1}{2} S_{ijkl} \sigma_{ij}^D \sigma_{kl}^D}_{\text{shear energy}} + \underbrace{\frac{1}{2} S_{ijkl} \sigma_H^2 \delta_{ij} \delta_{kl}}_{\text{hydrostatic energy}}.$$

When damage occurs in all directions,

$$w_e^* = \frac{1}{2} \frac{S_{ijkl}}{(1 - D_s)} \sigma_{ij}^D \sigma_{ij}^D + \frac{1}{2} \frac{S_{ijkl}}{(1 - D_n)} \sigma_H^2 \delta_{ij} \delta_{kl}.$$

This allows us to determine

— the law of elasticity, $\left\{ \begin{array}{lcl} \varepsilon_{ij}^{eD} & = & \dfrac{\partial w_e^*}{\partial \sigma_{ij}^D} \\ \varepsilon_H^e & = & \dfrac{\partial w_e^*}{\partial \sigma_H} \end{array} \right.$;

— the two variables associated to D_s and D_n, $\left\{ \begin{array}{lclcl} \overline{Y}_s & = & -\dfrac{\partial w_e^*}{\partial D_s} & \text{or } Y_s = & \dfrac{\partial w_e^*}{\partial D_s} \\ \overline{Y}_n & = & -\dfrac{\partial w_e^*}{\partial D_n} & \text{or } Y_n = & \dfrac{\partial w_e^*}{\partial D_n} \end{array} \right.$,

the damage dissipation being

$$Y_s \dot{D}_s + Y_n \dot{D}_n.$$

With this theory it is also possible to consider anisotropic damage for which $D_n \neq 0$ in certain material directions, and $D_n = 0$ in the nondamageable directions. To illustrate this point, let us develop the whole theory for an elementary ply of a fiber composite made of (O. Allix 1989)

— parallel fibers considered as linearly elastic, anisotropic and nondamageable up to brittle failure;
— a matrix whose behavior is governed by anisotropic linear elasticity coupled with damage by progressive cracking in the direction of the fibers;

— an interface which is linearly elastic and damageable in shear and elastic and brittle in transverse tension perpendicular to the fibers.

Assume plane stress conditions in the (x_1, x_2) plane. Take E_1 as Young's modulus in the x_1 fiber direction, E_2 as Young's modulus in the transverse x_2 direction, ν_{12} and ν_{21} as the corresponding Poisson's ratios, and G_{12} as the shear modulus. As the damage does not affect the behavior in the x_1 direction, must be equal to D_n in the x_2 direction and equal to D_s in (1,2) shear, then the complementary energy density must be written as

$$w_e^* = \frac{1}{2}\left[\frac{\sigma_{11}^2}{E_1} + \frac{\sigma_{22}^2}{E_2(1-D_n)} - \left(\frac{\nu_{12}}{E_1} + \frac{\nu_{21}}{E_2}\right)\sigma_{11}\sigma_{22} + \frac{\sigma_{12}^2}{G_{12}(1-D_s)}\right].$$

The plane strain components of the law of elasticity coupled with this particular damage are derived from this elastic stress potential (α and β being indices taking the values 1 and 2):

$$\varepsilon_{\alpha\beta} = \frac{\partial w_e^*}{\partial \sigma_{\alpha\beta}} \begin{cases} \varepsilon_{11} = \dfrac{\sigma_{11}}{E_1} - \left(\dfrac{\nu_{12}}{E_1} + \dfrac{\nu_{21}}{E_2}\right)\dfrac{\sigma_{22}}{2} \\ \varepsilon_{22} = \dfrac{\sigma_{22}}{E_2(1-D_n)} - \left(\dfrac{\nu_{12}}{E_1} + \dfrac{\nu_{21}}{E_2}\right)\dfrac{\sigma_{11}}{2} \\ \varepsilon_{12} = \dfrac{\sigma_{12}}{G_{12}(1-D_s)} \end{cases}.$$

The result is also obtained by homogenization techniques from the constitutive equations of the three components of the ply, namely the fibers, the matrix and the interface; it allows us to identify the damages D_n and D_s by experiments in tension σ_{22} in the x_2 direction and in shear σ_{12}.

The associated variables needed to model the kinetic laws of evolution of the damages are:

$$Y_s = \frac{\partial w_e^*}{\partial D_s} = \frac{\sigma_{12}^2}{2G_{12}(1-D_s)^2},$$

$$Y_n = \frac{\partial w_e^*}{\partial D_n} = \frac{\sigma_{22}^2}{2E_2(1-D_n)^2}.$$

2.2.4 Exercise on Anisotropic Damage in Proportional Loading

The case of proportional loading occurs quite often in practice. Considered as a special case, it is of great interest because it allows three-dimensional problems to be solved by working with scalar quantities only.

Definition: A loading on a particular structure is considered to be proportional when the stress field may be written as the product of a tensor which depends only upon the point considered, M, multiplied by a scalar which depends only upon the time:

$$\boldsymbol{\sigma}_{(M,t)} = \mathbf{S}_{(M)} \cdot \tau_{(t)}.$$

Demonstrate that the principal directions of the stresses remain constant with time (but may possibly be different at each point):

$$[\sigma] = \begin{bmatrix} S_1 & 0 & 0 \\ 0 & S_2 & 0 \\ 0 & 0 & S_3 \end{bmatrix} \cdot \tau_{(t)}.$$

Demonstrate that the triaxiality ratio remains constant with time:

$$S_H = \frac{1}{3} S_{kk},$$

$$S_{ij}^D = S_{ij} - S_H \delta_{ij} \quad S_{eq} = \left(\frac{3}{2} S_{ij}^D S_{ij}^D \right)^{1/2},$$

$$\boxed{\frac{\sigma_H}{\sigma_{eq}} = \frac{S_H}{S_{eq}} \text{ function of } M \text{ only}}.$$

Then the triaxiality function $R\nu$ also remains constant with time.

— Considering isotropic, linear elasticity coupled with anisotropic damage represented by the second order tensor D_{ij} of Section 2.2.1, *write the law of elasticity coupled with damage by means of the effective stress associated with the principle of strain equivalence,*

$$\frac{1}{2}\left[S_{ik}(1-D)_{kj}^{-1} + (1-D)_{ik}^{-1} S_{kj} \right] \tau_{(t)} = \frac{E}{1+\nu} \varepsilon_{ij}^e + \frac{\nu E}{(1+\nu)(1-2\nu)} \varepsilon_{kk}^e \delta_{ij},$$

and *show that the elastic strain tensor and the strain energy release rate tensor are also "proportional"*:

$$\varepsilon_{ij(M,t)}^e = e_{ij(M)} \tau_{(t)},$$

$$Y_{ij} = \frac{1}{2} a_{imkl} \varepsilon_{jm}^e \varepsilon_{kl}^e = Y_{ij(M)} \cdot \tau_{(t)}^2.$$

— Assuming that the damage rate depends linearly on the strain energy density release rate (see Chapter 3), *show that the damage tensor is also "proportional" for the initial condition* $D_{ij}(t = 0) = 0$:

$$\dot{D}_{ij} \sim Y_{ij} \rightarrow D_{ij(M,t)} = \Delta_{ij(M)} D_{(t)},$$

where $D_{(t)}$ is a scalar function of time.

— Finally, *demonstrate that the damage has the same principal directions as the stress*, allowing us to write the effective stress matrix as (see Section 2.2.1)

$$[\tilde{\sigma}] = \begin{bmatrix} \dfrac{\sigma_1}{1-D_1} & 0 & 0 \\ 0 & \dfrac{\sigma_2}{1-D_2} & 0 \\ 0 & 0 & \dfrac{\sigma_3}{1-D_3} \end{bmatrix}.$$

These results are very important for the calculation of the damage evolution in all applications satisfying the conditions of proportional loading. The tensor $\mathbf{\Delta}$ has to be calculated only once from the reduced stress $\mathbf{S}$, and the further evolution depends only upon the scalar D through a straightforward integration of the kinetic law of damage described in Chapter 3.

2.3 Micromechanics of Damage

Micromechanics consists of deriving the behavior of materials at the mesoscale from the study of specific mechanisms at the microscale. The mechanisms must be well defined from physical observations by precise geometries and kinematics. Their mechanical modeling is performed with elementary, common constitutive equations for strain, crack growth and fracture known at the meso or macroscale. The interest comes from the interaction between mechanisms and from the homogenization brindging the gap between micro- and mesoscales.

2.3.1 Brittle Isotropic Damage

Microcracks and the damage variable

The main mechanism of brittle damage is the nucleation and growth of microcracks and their coalescence to initiate a crack at the mesoscale. Let us derive from the microcrack pattern the relation between the isotropic damage variable D defined at the mesoscale, and the surface density of microcracks.
Consider for simplicity a R.V.E. at the mesoscale as a cube of dimension $l \cdot l \cdot l$ and define the microscale of cells as the cube $d \cdot d \cdot d$ in which a microcrack of any area s_i and any orientation may lie (Figure 2.11). The number of cells is l^3/d^3 although the number of cracks, is $n (n \leq l^3/d^3)$.
The geometry having been defined, modeling consists here in writing the balance of the dissipated energy calculated by classical fracture mechanics and the dissipated energy calculated by continuous damage mechanics.

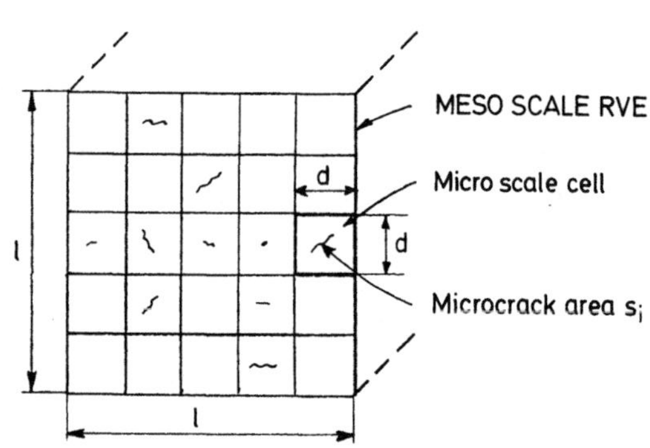

Fig. 2.11. Micro-meso element for brittle damage

If G is the strain energy release rate corresponding to a crack of area s, D the equivalent damage of the cell and Y the strain energy density release rate for one cracked cell loaded by a given state of stress ;

$$G\dot{s} = Y\dot{D}d^3.$$

For the n cracked cells of the mesocube :

$$\sum_1^n G_i\dot{s}_i = \sum_1^n Y_i\dot{D}_i d^3.$$

Assuming that brittle growth of microcracks occurs at $G = G_c = \text{const.}$, and that brittle damage occurs at $Y = Y_c = \text{const.}$,

$$G_c\sum_1^n \dot{s}_i = Y_c d^3\sum_1^n \dot{D}_i.$$

Furthermore, when $s_i = 0 \to D_i = 0$, the obvious integration yields

$$G_c\sum_1^n s_i = Y_c d^3\sum_1^n D_i.$$

The simplest homogenization consists in defining the damage D at the mesoscale by the mean value of the damages of all the cells, whether cracked or not. Their number is l^3/d^3:

$$D = \frac{d^3}{l^3}\sum_1^n D_i$$

or

$$D = \frac{G_c}{Y_c l^3}\sum_1^n s_i.$$

The term $G_c/Y_c l^3$ may be calculated from the rupture criterion for mesocrack initiation in one plane or here in a set of cells occupying the flat volume $l \cdot l \cdot d$. Due to the localization of damage, other microcracks may be neglected.
If one assumes that the mesocrack is initiated when a part k of this flat volume has microcracks of size d^2, corresponding to $D = D_c$ at the mesoscale, then

$$\sum_1^n s_i = kl^2 \quad D_c = \frac{G_c}{Y_c l^3}kl^2 \text{ or } \frac{G_c}{Y_c l^3} = \frac{D_c}{kl^2}.$$

This allows us to write D as

$$D = \frac{\sum_1^n s_i}{l^2}\frac{D_c}{k},$$

which shows that D is the surface density of microcracks multiplied by a correction factor (here, D_c/k) as was mentioned for the definition of the damage variable in Section 1.2.1.
If the simplest fracture criterion is considered,

$$\sum_1^n s_i = l^2 \to k = 1 \to D_c = 1,$$

then, the damage is simply

$$D = \frac{\sum_1^n s_i}{l^2}.$$

This result was also found by the purely geometrical considerations described in Section 1.3.1.
By the way, this calculation gives the order of magnitude of a characteristic length which permits correlation between fracture mechanics and damage mechanics, namely l, the size of the R.V.E.
Taking the simple fracture criterion $k = 1$, $D_c = 1$,

$$\frac{G_c}{Y_c l^3} = \frac{1}{l^2};$$

then

$$\boxed{l \simeq \frac{G_c}{Y_c}}$$

for most metallic materials from the table in Section 3.5:

light alloys		steels and alloys
0.005	$\le G_c \le$	0.05 MPa.m
2	$\le Y_c \le$	10 MPa
$2.5\ 10^{-3}$	$\le l \le$	$5\ 10^{-3}$ m

and for concrete in tension: $G_c \simeq 3\ 10^{-5}$ MPa.m, $Y_c \simeq 1.5\ 10^{-4}$ MPa: $l \simeq 2\ 10^{-1}$ m.
This shows that the size of the physical RVE must be of the order of millimeters for metals and of the order of decimeters for concrete as pointed out in Section 1.1.3, where slightly smaller values were involved.

Micromechanics and damage growth

As a specific example, let us derive the kinetic damage evolution law at the mesoscale which corresponds to fatigue microcrack growth at the microscale of Figure 2.11.
With $D_c = k$, the damage was found to be

$$D = \frac{\sum_1^n s_i}{l^2}$$

or, for its rate,

$$\dot{D} = \frac{\sum_1^n \dot{s}_i}{l^2}.$$

The surface growth rate of one crack $\dot{s}_i$ may be expressed as a function of the strain energy release rate of one cell G_i by means of the Paris law of fatigue crack growth.

If, for a two-dimensional problem,

— $2a$ is the crack length, $\dfrac{\delta a}{\delta N}$ is the crack increment per cycle,
— N the number of cycles of loading in mode I,
— and K_M the amplitude of the stress intensity factor (with $K_{\min} = 0$), then

$$\frac{\delta a}{\delta N} = CK_M^{\eta},$$

where C and η are two material parameters, $\eta \simeq 4$ for many metallic materials.

Paris' relation may be considered as the integration over one cycle of the rate equation:

$$\dot{a} = \eta C K^{\eta-1}\dot{K}.$$

Or, for the cell problem with the relations

$$K = (EG)^{1/2}, \quad \dot{K} = \frac{E^{1/2}}{2}G^{-1/2}\dot{G}, \quad \dot{s}_i = 2e_i\dot{a}_i,$$

where e_i is the width of the crack (i),

$$\dot{s}_i = \eta C e_i E^{\eta/2} G_i^{\eta/2-1}\dot{G}_i.$$

A relation between G_i and Y_i may be found through their definitions by the elastic energy. If w_{ei} is the elastic strain energy density and W_{ei} the elastic strain energy of the elementary cell,

$$G_i = \frac{\partial W_{ei}}{\partial s_i}, \quad \overline{Y}_i = \frac{\partial w_{ei}}{\partial D_i}, \quad Y_i = -\overline{Y}_i = -\frac{\partial w_{ei}}{\partial D_i},$$

with

$$W_{ei} = w_{ei}d^3;$$

$$G_i = -\frac{\partial(w_{ei}d^3)}{\partial D_i}\frac{dD_i}{ds_i},$$

with

$$D_i \simeq \frac{s_i}{d^2};$$

$$G_i = Y_i d, \quad \dot{G}_i = \dot{Y}_i d,$$

and

$$\dot{s}_i = \eta C e_i E^{\eta/2} d^{\eta/2} Y_i^{\eta/2-1}\dot{Y}_i.$$

The damage rate is

$$\dot{D} = \frac{\Sigma \dot{s}_i}{l^2} = \frac{\eta C E^{\eta/2} d^{\eta/2}}{l^2}\sum_1^n e_i Y_i^{\eta/2-1}\dot{Y}_i.$$

Assuming that all the n cracked cells have the same strain energy density release rate $Y_i = Y_n$, the homogenized strain energy density release rate for the R.V.E. is

$$Y = nY_n \text{ and also } \dot{Y} = n\dot{Y}_n.$$

Also assuming the same width for all the cracks $e_i = e$,

$$\sum_1^n e_i Y_i^{\frac{\eta}{2}-1}\dot{Y}_i = neY_n^{\frac{\eta}{2}-1}\dot{Y}_n = en^{1-\frac{\eta}{2}}Y^{\frac{\eta}{2}-1}\dot{Y},$$

$$\dot{D} = \frac{\eta C E^{\eta/2} d^{\eta/2} e}{l^2 n^{\frac{\eta}{2}-1}} Y^{\frac{\eta}{2}-1}\dot{Y}.$$

This example shows that the damage rate is an increasing function of the strain energy density release rate and is quasi-proportional to it for most materials for which $\eta \simeq 4$. It is also proportional to its rate $\dot{Y}$. This fact will be used as a guideline for the derivation of a general kinetic law of damage evolution in Chapter 3.

2.3.2 Ductile Isotropic Damage

Microcavities and the damage variable

The main mechanisms of ductile damage are the nucleation, growth and coalescence of microcavities by large local plastic deformations. As in the previous section, let us derive a relation between the isotropic damage variable D at the mesoscale and the density of microvoids.

Let us again consider a RVE at the mesoscale as a cube $l \cdot l \cdot l$ with n cavity cells of dimension d^3 (Figure 2.12).

On this very simple geometry let us again write the balance of energy calculated from the growth of cavities and from the damage concept.

According to the Gurson model, the porosity at the mesoscale P is equal to the hydrostatic part of the plastic strain $\varepsilon_H^p = \frac{1}{3}\varepsilon_{kk}^p$ due to the growth of voids. For the geometrical model under consideration,

$$P = \frac{nd^3}{l^3},$$

where n is the number of cavities.

Writing the equality for the rates at the mesoscale,

$$\dot{P} = \dot{\varepsilon}_H^p$$

The total power density dissipated at the mesoscale for the homogenized stress σ_{ij} and plastic strain rate $\dot{\varepsilon}_{ij}^p$ is

$$\sigma_{ij}\dot{\varepsilon}_{ij}^p,$$

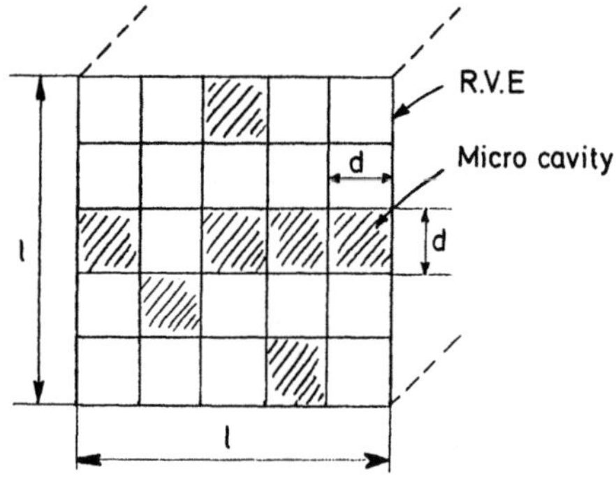

Fig. 2.12. Micro-meso element for ductile damage

which can be split into two parts by means of the deviatoric and the hydrostatic quantities:

$$\left(\sigma_{ij}^D + \sigma_H \delta_{ij}\right)\left(\dot{\varepsilon}_{ij}^{pD} + \dot{\varepsilon}_H^p \delta_{ij}\right)$$

or

$$\sigma_{ij}^D \dot{\varepsilon}_{ij}^{pD} + 3\sigma_H \dot{\varepsilon}_H^p + 0.$$

The first term is the power dissipated in pure plasticity by slips. The second term, which corresponds to the irreversible change of volume, may be interpreted as the power dissipated for the increase of material discontinuities in the RVE by the growth of cavities. It is this part which has to be equal to the damage dissipation,

$$3\sigma_H \dot{\varepsilon}_H^p l^3 = Y \dot{D} l^3$$

or

$$\dot{D} = \frac{3\sigma_H}{Y} \dot{P}.$$

For simplicity, let us assume proportional loading, perfect plasticity, $\sigma_H / Y = \text{const}$ and the initial condition $P = 0 \rightarrow D = 0$; then, integration yields

$$D = \frac{3\sigma_H}{Y} n \frac{d^3}{l^3}.$$

As for the brittle damage, it is assumed that the mesocrack initiation criterion is reached when a set of cavities fulfill a flat volume of dimension $l \cdot l \cdot d$, the other cavities in the RVE being neglected because of the the localization phenomenon. The critical value of the porosity corresponding to $D = 1$ is

$$P_c = \frac{l^2 d}{l^3} = \frac{d}{l},$$

which allows for the calculation of the term $3\sigma_H / Y$ in the damage equation

$$1 = \frac{3\sigma_H}{Y} \frac{d}{l}.$$

Finally, we obtain the equation

$$D = n \frac{d^2}{l^2},$$

which was introduced in Section 1.2.1.

Microcavities and damage growth

The kinetic law of damage evolution may be derived from the above expression for D:

$$\dot{D} = \frac{d^2}{l^2} \dot{n} + 2n \frac{d \dot{d}}{l^2}.$$

The first term accounts for the increase in the number of cavities ($\dot{n}$ signifies the number of cavities per unit time), and the second term accounts for the cavity growth.

a) Growth by nucleation of the number of cavities:

$$\dot{D} = \frac{d^2}{l^2} \dot{n}.$$

In the Gurson model, the porosity rate is also the sum of two terms accounting for nucleation and growth. For nucleation the Tvergaard kinetic law is used:

$$\dot{P} = A\dot{\sigma}_{eq} + B\dot{\sigma}_H,$$

where A and B are material parameters.
Assuming for simplicity a sudden nucleation of cavities of a fixed size d,

$$\dot{P} = \dot{n}\frac{d^3}{l^3},$$

$$\dot{D} = \dot{P}\frac{l}{d} = \frac{l}{d}\left(A\dot{\sigma}_{eq} + B\dot{\sigma}_H\right)$$

or

$$\dot{D} = \frac{l}{d}\dot{\sigma}_{eq}\left(A + B\frac{\dot{\sigma}_H}{\dot{\sigma}_{eq}}\right)$$

It is convenient to express the damage as a function of the accumulated plastic strain rate $\dot{p} = \left(\frac{2}{3}\dot{\varepsilon}^p_{ij}\dot{\varepsilon}^p_{ij}\right)^{1/2}$, which is easy to introduce by means of the plastic tangent modulus E_T.
Assuming proportional loading,

$$\dot{p} = \frac{\dot{\sigma}_{eq}}{E_T};$$

and also

$$\frac{\dot{\sigma}_H}{\dot{\sigma}_{eq}} = \frac{\sigma_H}{\sigma_{eq}},$$

$$\dot{D} = \frac{l}{d}E_T\left(A + B\frac{\sigma_H}{\sigma_{eq}}\right)\dot{p}.$$

b) Growth by enlargement of cavities of a fixed number n:

$$\dot{D} = 2n\frac{d\dot{d}}{l^2}$$

or with

$$D = n\frac{d^2}{l^2},$$

$$\dot{D} = 2D\frac{\dot{d}}{d}.$$

The problem of crack growth has received much attention in the past 20 years. An essential result for this problem is offered by the McClintock and Rice and Tracey analysis, which derives the rate of growth of a cylindrical or a spherical cavity of volume V in a perfectly plastic infinite body as a function of the accumulated plastic strain rate $\dot{p}$ and the triaxiality ratio σ_H/σ_{eq}:

$$\dot{V} = 0.85V\dot{p}\exp\left(\frac{3}{2}\frac{\sigma_H}{\sigma_{eq}}\right)$$

or, with

$$V = d^3,$$

$$3d^2\dot{d} = 0.85d^3\dot{p}\exp\left(\frac{3}{2}\frac{\sigma_H}{\sigma_{eq}}\right),$$

$$\dot{D} = 0.57 D \dot{p} \exp\left(\frac{3}{2}\frac{\sigma_H}{\sigma_{eq}}\right).$$

These two examples of cavity nucleation and growth show that the damage rate is

— proportional to the accumulated plastic strain rate ;
— an increasing function of the triaxiality ratio σ_H/σ_{eq};
— a function of the state through E_T or D.

These properties will also serve as a guideline for the derivation a general kinetic law of damage evolution in Chapter 3.

2.3.3 Anisotropic Damage (D. Krajcinovic 1981)

A way to modelling anisotropic damage is to consider its influence on the stiffness or on the compliance of the material at the mesoscale, that is, to generalize the isotropic elastic strain energy density of a damaged material to the anisotropic case:

$$w_e = \frac{1}{2}\tilde{a}_{ijkl}\varepsilon^e_{ij}\varepsilon^e_{kl},$$

in which the damage D, here isotropic, is defined as

$$\tilde{a}_{ijkl} = a_{ijkl}(1 - D).$$

Analysis of a single crack cell

Consider a cell at the microscale containing a circular crack defined by Euler's angles θ, ϕ of the orientation of its plane, by its radius a and its area A (Figure 2.13). The matrix is isotropic linear elastic and the response is perfectly brittle.

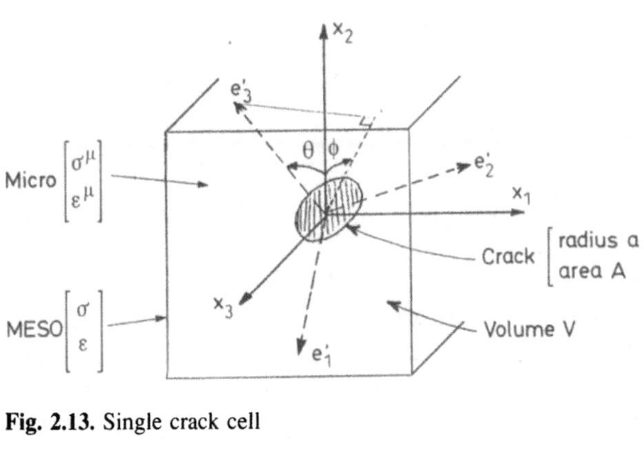

Fig. 2.13. Single crack cell

Let σ^{μ} and ε^{μ} be the stress and elastic strain tensor fields in the cell, decomposed in

$$\begin{aligned}\sigma^{\mu} &= \sigma^{o} + \sigma^{c},\\ \varepsilon^{\mu} &= \varepsilon^{o} + \varepsilon^{c},\end{aligned}$$

where σ^{o} and ε^{o} are the stress and elastic strain "far fields", which are constant in the cell when no crack is involved, and σ^{c} and ε^{c} are the disturbances in stress and strain due to the crack. These fields verify the classical equations of continuum mechanics, assuming a static case and no body forces:

$$\begin{aligned}\sigma^{o}_{ij,j} &= 0,\\ \sigma^{o}_{ij}n_j &= F_i,\\ \varepsilon^{o}_{ij} &= S^{o}_{ijkl}\sigma^{o}_{kl},\end{aligned}$$

where $\mathbf{S}^{o}$ is the elastic compliance tensor of the matrix.
The total stress and strain also verify that

$$\varepsilon^{o}_{ij} + \varepsilon^{c}_{ij} = S^{o}_{ijkl}\left(\sigma^{o}_{kl} + \sigma^{c}_{kl}\right),$$

and the stress disturbance verify that

$$\begin{aligned}\sigma^{c}_{ij,j} &= 0,\\ \sigma^{c}_{ij}n_j &= 0,\end{aligned}$$

since σ^{o} is considered to be a "far field" for the crack.
The conditions of free surfaces are written on the two surfaces of a crack of area A as

$$\left(\sigma^{o}_{ij} + \sigma^{c}_{ij}\right) n_j = 0 \text{ on } A.$$

The important role in the loss of stiffness of the cell due to the crack is played by the crack opening displacement vector $\vec{b}$ defined by the discontinuity of the displacement across the opened crack,

$$b_i = [[u^{o}_i + u_i]].$$

The relation existing between $\vec{b}$ and the "far field" stress $\boldsymbol{\sigma}^{o}$ is somewhat difficult to obtain. Using Eshelby's result for an ellipsoidal inclusion of vanishing thickness and Mura's computations, the result with respect to the crack coordinates, where e'_3 is normal to the crack in its center, is

$$b'_i = \left[1 - \left(\frac{x'_j}{a}\right)^2\right]^{1/2} B'_{ik}\sigma^{o'}_{3k}a,$$

where $x'_j < a$ is measured along the coordinate axis e'_j and B'_{ik} is a symmetric matrix of influence coefficients.
This equation represents the ellipsoidal void created by the opening of a circular plane crack. It allows for the determination of all the stress and strain fields in the cell. "This involves arduous computations using Eshelby tensor and elliptic integrals" (D. Krajcinovic). In fact, to characterize the mesoresponse of the cell it is not necessary to perform these calculations because it is governed by the averages of the microstress and -strain fields. The local fluctuations are of a second order nature.

The total stress averaged over the volume of the unit cell is

$$\sigma_{ij} = \frac{1}{V}\int\limits_V \left(\sigma_{ij}^o + \sigma_{ij}^c\right) dV = \sigma_{ij}^o + \frac{1}{V}\int\limits_V \sigma_{ij}^c \, dV.$$

To perform the integration by parts let us write the equation together with $x_{j,k} = \delta_{jk}$ as

$$\sigma_{ij} = \sigma_{ij}^o + \frac{1}{V}\int\limits_V \sigma_{ik}^c x_{j,k} \, dV$$

or

$$\sigma_{ij} = \sigma_{ij}^o + \frac{1}{V}\left[\int\limits_A \sigma_{ik}^c n_k x_j \, dA - \int\limits_V \sigma_{ik,k}^c x_j \, dV\right].$$

The equilibrium equations for the stress disturbance require the two integrals to vanish and, as expected ; we obtain

$$\sigma_{ij} = \sigma_{ij}^o.$$

The average of the strain may be deduced from the stress and from the contribution of the crack-opening displacement:

$$\varepsilon_{ij} = \frac{1}{V}\int\limits_V (\varepsilon_{ij}^o + \varepsilon_{ij}^c) \, dV = S_{ijkl}^o \frac{1}{V}\int\limits_V (\sigma_{kl}^o + \sigma_{kl}^c) \, dV + \varepsilon_{ij}^*,$$

$$\varepsilon_{ij} = S_{ijkl}^o \sigma_{kl} + \varepsilon_{ij}^*.$$

The determination of ε_{ij}^* first requires the calculation of the average of the crack-opening displacement over the crack area:

$$\int\limits_A b_i' \, dA;$$

then, the transformation from local (e_1', e_2', e_3') to global (x_1, x_2, x_3) coordinates by means of the rotation matrix g_{ij}:

$$e_i' = g_{ij} x_j,$$

from which it follows that

$$b_i' = g_{ij} b_j \text{ and } \sigma_{ij}' = g_{ik} g_{jl} \sigma_{kl}.$$

In concise form, the result for tension is

$$\varepsilon_{ij}^* = a^3 b_{ijkl} \sigma_{kl},$$

where the fourth order tensor **b** has the form,

$$b_{ijkl} = \frac{\pi}{3} B_{mn} (n_j g_{mi} + n_i g_{mj})(g_{nl} - f \delta_{2n} n_l) n_k,$$

in which the nonvanishing components of **B** are

$$\begin{bmatrix} B_{11} \\ B_{22} \\ B_{33} \end{bmatrix} = \frac{8(1-\nu^2)}{\pi E(2-\nu)} \begin{bmatrix} 2 \\ 2 \\ (2-\nu) \end{bmatrix}.$$

E and ν are Young's modulus and Poisson's ratio, and f is Coulomb's coefficient of sliding friction between the two surfaces of the crack.

The final expression for the homogenized stress strain relation in a single crack cell loaded in tension is

$$\varepsilon_{ij} = \frac{1+\nu}{E}\sigma_{ij} - \frac{\nu}{E}\sigma_{kk}\delta_{ij} + \frac{a^3}{V}\frac{8(1-\nu^2)}{3E(2-\nu)}\begin{pmatrix} \delta_{ik}n_j n_l \sigma_{kl} \\ +\delta_{il}n_j n_k \sigma_{kl} \\ +\delta_{jk}n_l n_i \sigma_{kl} \\ +\delta_{jl}n_i n_k \sigma_{kl} \\ -2\nu n_i n_j n_k n_l \sigma_{kl} \end{pmatrix}.$$

Analysis of an anisotropic distribution of cracks

Consider now a set of cells each containing a crack of different orientation. Let us calculate the strain energy of the whole.
According to Taylor's hypothesis we consider a reasonably dilute concentration of microcracks which allows us to neglect the interactions between the cracks. The above analysis may be used considering a random distribution of orientations and sizes of the crack (Figure 2.14). The strain energy is simply the superposition of the contribution of each cell.
For one active crack cell, the strain energy density is calculated from

$$w_e = \frac{1}{2}\sigma_{ij}\varepsilon_{ij},$$

with ε_{ij} taken from the previous expression.
In order to obtain w_e as a function of the strain, a Fenchel transform is performed:

$$w_e = \mu\varepsilon_{ij}\varepsilon_{ij} + \frac{\lambda}{2}\varepsilon_{kk}\varepsilon_{ll} - \frac{32}{3}\frac{(\lambda+2\mu)\mu}{3\lambda+4\mu}\frac{a^3}{V}\begin{pmatrix} \frac{1-\nu}{8}\left(\frac{\lambda}{\mu}\right)^2 \varepsilon_{kk}\varepsilon_{ll} \\ +\varepsilon_{ik}\varepsilon_{kj}n_i n_j + \frac{\lambda(2-\nu)}{2\mu}\varepsilon_{kk}\varepsilon_{ij}n_i n_j \\ -\frac{\nu}{2}(\varepsilon_{ij}n_i n_j)(\varepsilon_{kl}n_k n_l) \end{pmatrix},$$

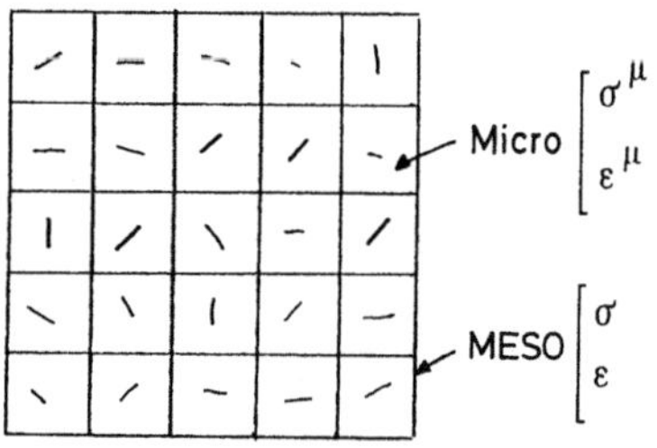

Fig. 2.14. Multiple crack cell element

where λ and μ are Lamé's constants:

$$\lambda = \frac{\nu E}{(1+\nu)(1-2\nu)}, \quad \mu = \frac{E}{2(1+\nu)}.$$

The measure of the damage is here $\frac{a^3}{V}$, its relation with the D variable used all along the course is obvious.
Considering a cubic cell of length l : $V = l^3$,

$$D_{(\vec{e}'_3)} = \frac{\pi a^2}{l^2}$$

and

$$\frac{a^3}{V} = \left(\frac{D}{\pi}\right)^{3/2}.$$

If we now consider a set of cells with a random distribution of cracks, the calculation of the mean value of the energy requires us to consider the following terms as random variables:

$$\frac{a^3}{V},$$
$$\frac{a^3}{V} n_i n_j,$$
$$\frac{a^3}{V} n_i n_j n_k n_l.$$

Calling $p\left(\frac{a^3}{V}, \vec{n}\right)$ the probability density associated with the relative void area of the cracks $\frac{a^3}{V}$ and their orientations $\vec{n}$, let us introduce the Leckie and Onat tensors:

$$\omega = \frac{1}{V} \int_V \frac{a^3}{V} p_{(a^3/V,\vec{n})} \, dV,$$

$$\omega_{ij} = \frac{1}{V} \int_V \frac{a^3}{V} p_{(a^3/V,\vec{n})} n_i n_j \, dV,$$

$$\omega_{ijkl} = \frac{1}{V} \int_V \frac{a^3}{V} p_{(a^3/V,\vec{n})} n_i n_j n_k n_l \, dV.$$

Then, the elastic strain energy density of an anisotropic mesoelement containing cracks defined by the probability density of their relative size and orientation $p\left(\frac{a^3}{V}, \vec{n}\right)$ is

$$w_e = \frac{1}{V} \int w_e \left(\frac{a^3}{V}, \vec{n}\right) dV,$$

$$w_e = \mu\varepsilon_{ij}\varepsilon_{ij} + \frac{\lambda}{2}\varepsilon_{kk}\varepsilon_{ll} - \frac{32}{3}\frac{(\lambda+2\mu)\mu}{3\lambda+4\mu}\begin{pmatrix} \omega\frac{1-\nu}{8}\left(\frac{\lambda}{\mu}\right)^2\varepsilon_{kk}\varepsilon_{ll} \\ +\omega_{ij}\left(\varepsilon_{ik}\varepsilon_{kj} + \frac{\lambda(2-\nu)}{2\mu}\right)\varepsilon_{kk}\varepsilon_{ij} \\ -\omega_{ijkl}\frac{\nu}{2}\varepsilon_{ij}\varepsilon_{kl} \end{pmatrix}.$$

By analogy with the isotropic case, this expression shows that the description of the anisotropic damage is achieved by three variables:

— a scalar ω as in the isotropic case ;
— a second-order tensor ω_{ij} as may be obtained by a purely geometric analysis (see Section 2.2.1);
— a fourth-order tensor ω_{ijkl} as it is obtained by a pure thermodynamical analysis (see Section 2.2.2).

2.3.4 Microcrack Closure Effect, Unilateral Conditions

Derivation of a crack closure parameter

In Section 1.2.1 the effective stress in tension was defined as:

$$\tilde{\sigma} = \frac{\sigma}{1-D}.$$

This is also the effective stress in compression if the microcracks and microcavities remain open. For certain materials and certain conditions of loading, the defects may close in compression. This is often the case for very brittle materials. If the defects close completely, the area which effectively carries the load in compression is no longer $(S - S_D)$, but S. This means that the effective area is defined by a unilateral condition :

$$\tilde{S}^+ = S - S_D \text{ in tension } \sigma^+ \rightarrow \tilde{S}^+ = S(1-D) \rightarrow \tilde{\sigma}^+ = \frac{\sigma^+}{1-D},$$
$$\tilde{S}^- = S \text{ in compression } \sigma^- \rightarrow \tilde{\sigma}^- = \sigma^-.$$

In fact, the real defects of complicated shapes do not close completely. The effective area in compression is such that

$$S - S_D < \tilde{S} < S.$$

Let us write this expression as $\tilde{S} = S - hS_D = S(1-Dh)$, where h, $(0 \leq h \leq 1)$ is a crack closure parameter which depends a priori upon the material and thc loading. Let us determine h from a very simple micromechanical model. Consider again a cubic R.V.E. of size $l \cdot l \cdot l$ containing a plane crack of area s loaded in mode I by a force F with a damaged zone created by the growth of the crack (Figure 2.15).
The material is assumed to be linearly elastic with a Young's modulus E. It is also elastic in the damage zone with an elasticity modulus $\tilde{E} = E(1-D_c)$, where D_c is the critical value of the damage during the initiation and growth of the crack.

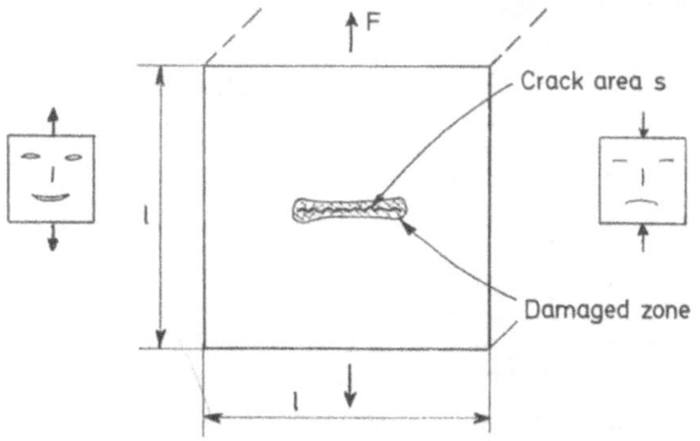

Fig. 2.15. Microcrack in quasi-unilateral condition

Let us write the overall equilibrium equations in the plane of the crack.

— In tension, $F = F^+$, the area of resistance is $(l^2 - s)$:

$$F^+ = \int_{l^2-s} \sigma^+_{(M)}\, dS = \int_{l^2-s} E\varepsilon^+_{(M)}\, dS.$$

For a rough approximation we may consider the Lin-Taylor hypothesis, which states that the strain is uniform in the plane of the defect; then

$$F^+ = E\varepsilon^+(l^2 - s).$$

The homogenization at the mesoscale consists in writing the mesostress as $\sigma^+ = \dfrac{F^+}{l^2}$:

$$\sigma^+ = E\varepsilon^+\left(1 - \frac{s}{l^2}\right)$$

or with the simple definition of damage $D = \dfrac{s}{l^2}$ and the principle of strain equivalence,

$$\sigma^+ = E\varepsilon^+(1 - D), \quad \tilde{\sigma} = \frac{\sigma^+}{1 - D}.$$

— In compression, $F = F^-$, the damage as a state variable is still D but its effect is different. The resisting area l^2 may be decomposed in two parts: $(l^2 - s)$, for which the elasticity modulus is E; and s, for which the elasticity modulus is one of the damaged zone with the crack closed $\tilde{E} = E(1 - D_c)$. Then the equilibrium equation is written as

$$F^- = \int_{l^2-s} \sigma^-_{(M)}\, dS + \int_s \sigma^-_{(M)}\, dS$$

or, with the same Lin-Taylor hypothesis as above,

$$F^- = E\varepsilon^-(l^2 - s) + E(1 - D_c)\varepsilon^- s,$$

which gives for the homogenized mesostress:

$$\sigma^- = \frac{F^-}{l^2} = E\varepsilon^-(1-D) + E(1-D_c)\varepsilon^- D,$$

$$\sigma^- = E\varepsilon^-[1-D+(1-D_c)D],$$

$$\sigma^- = E\varepsilon^-(1-DD_c).$$

The principle of strain equivalence requires that $\tilde{\sigma} = E\varepsilon^-$ and

$$\tilde{\sigma} = \frac{\sigma}{(1-D\cdot D_c)}.$$

This simple example shows that the crack closure parameter h is of the order of magnitude of D_c, a quantity which depends upon the material and the loading:

$$\boxed{h \simeq D_c}.$$

In practice, h is considered to be a constant, in order to avoid a tremendous amount of algebra! A value of $h = 0.2$ often gives results close to the experiments. It is also possible to identify its value from the measurements of elasticity modulus in tension and in compression on a damaged material as explained in Section 3.1.6.

Effective stress for the three-dimensional case

To define tension or compression by the sign of a scalar quantity in the one-dimensional case is easy. In three dimensions, speaking of the sign of a tensor or a matrix is like speaking of one color of a flag! A norm must be defined in accordance with the basic features of damage mechanics.
D is a state variable, it has the same value for tension and compression. Then the only problem remaining is to define the effective stress, which will be done under the following scheme.

— Decomposition of the stress tensor in a positive part and a negative part, from the signs of the principal stresses σ_i:

$$[\sigma] = \begin{bmatrix} \sigma_1 & 0 & 0 \\ 0 & \sigma_2 & 0 \\ 0 & 0 & \sigma_3 \end{bmatrix},$$

or with the Macauley brackets $\begin{cases} \langle x \rangle = x \text{ if } x \geq 0 \\ \langle x \rangle = 0 \text{ if } x < 0 \end{cases}$

$$[\sigma] = \begin{bmatrix} \langle\sigma_1\rangle & 0 & 0 \\ 0 & \langle\sigma_2\rangle & 0 \\ 0 & 0 & \langle\sigma_3\rangle \end{bmatrix} - \begin{bmatrix} \langle-\sigma_1\rangle & 0 & 0 \\ 0 & \langle-\sigma_2\rangle & 0 \\ 0 & 0 & \langle-\sigma_3\rangle \end{bmatrix},$$

denoted as $$\sigma_{ij} = \langle\sigma_{ij}\rangle - \langle-\sigma_{ij}\rangle$$

or $$\boldsymbol{\sigma} = \langle\boldsymbol{\sigma}\rangle - \langle-\boldsymbol{\sigma}\rangle.$$

Using the property $\langle x\rangle \cdot \langle -x\rangle = 0$ it is easy to demonstrate that

$$\sigma_{kk} = \langle \sigma_{kk}\rangle - \langle -\sigma_{kk}\rangle,$$
$$\sigma_{kk}^2 = \langle \sigma_{kk}\rangle^2 + \langle -\sigma_{kk}\rangle^2,$$
$$\sigma_{ij}\sigma_{ij} = \langle \sigma_{ij}\rangle\langle \sigma_{ij}\rangle + \langle -\sigma_{ij}\rangle\langle -\sigma_{ij}\rangle.$$

— Partition of the specific enthalpy in two terms which are functions of $\langle \boldsymbol{\sigma}\rangle$ and $\langle -\boldsymbol{\sigma}\rangle$. Instead of deriving the law of elasticity from the state potential specific free energy $\psi_e(\varepsilon^e)$ as in Section 2.1.1, it is more convenient here to work with its dual transform: the Gibbs specific enthalpy $\psi_e^*(\boldsymbol{\sigma})$.

For linear elasticity without any damage,

$$\psi_{e(\varepsilon^e)} = \frac{E}{2\rho}\left[\frac{1}{1+\nu}\varepsilon_{ij}^e\varepsilon_{ij}^e + \frac{\nu}{(1+\nu)(1-2\nu)}\varepsilon_{kk}^{e2}\right]$$
$$\rightarrow \psi_{e(\sigma)}^* = \frac{1}{2\rho E}[(1+\nu)\sigma_{ij}\sigma_{ij} - \nu\sigma_{kk}^2].$$

ψ_e^* may be written as a function of $\langle \sigma_{ij}\rangle$ and $\langle -\sigma_{ij}\rangle$:

$$\psi_e^* = \frac{1}{2\rho E}\left[(1+\nu)\langle \sigma_{ij}\rangle\langle \sigma_{ij}\rangle - \nu\langle \sigma_{kk}\rangle^2\right] +$$
$$+\frac{1}{2\rho E}\left[(1+\nu)\langle -\sigma_{ij}\rangle\langle -\sigma_{ij}\rangle - \nu\langle -\sigma_{kk}\rangle^2\right]$$

When isotropic damage occurs, in accordance with the concept of effective stress, the unilateral conditions are written by means of $(1-D)$ for the first term, and $(1-Dh)$ for the second term:

$$\psi_e^* = \frac{1}{2\rho E(1-D)}\left[(1+\nu)\langle \sigma_{ij}\rangle\langle \sigma_{ij}\rangle - \nu\langle \sigma_{kk}\rangle^2\right] +$$
$$+\frac{1}{2\rho E(1-Dh)}\left[(1+\nu)\langle -\sigma_{ij}\rangle\langle -\sigma_{ij}\rangle - \nu\langle -\sigma_{kk}\rangle^2\right].$$

From this, the law of elasticity is derived as

$$\varepsilon_{ij}^e = \rho\frac{\partial \psi_e^*}{\partial \sigma_{ij}},$$

with the properties demonstrated by P. Ladeveze (1984):

$$\frac{\partial}{\partial \sigma_{ij}}\left(\frac{1}{2}\langle \sigma_{ij}\rangle\langle \sigma_{ij}\rangle\right) = \frac{\partial}{\partial \sigma_{ij}}\left(\frac{1}{2}\langle \sigma_{ik}\rangle\langle \sigma_{kj}\rangle\delta_{ij}\right) = \langle \sigma_{ij}\rangle,$$
$$\frac{\partial}{\partial \sigma_{ij}}\left(\frac{1}{2}\langle \sigma_{kk}\rangle^2\right) = \frac{\partial}{\partial \sigma_{ij}}\left(\frac{1}{2}\langle \sigma_{kk}\rangle\langle \sigma_{kk}\rangle\right) = \langle \sigma_{kk}\rangle\delta_{ij},$$
$$\varepsilon_{ij}^e = \frac{1+\nu}{E(1-D)}\langle \sigma_{ij}\rangle - \frac{\nu}{E(1-D)}\langle \sigma_{kk}\rangle\delta_{ij} - \frac{1+\nu}{E(1-Dh)}\langle -\sigma_{ij}\rangle$$
$$+\frac{\nu}{E(1-Dh)}\langle -\sigma_{kk}\rangle\delta_{ij}.$$

— Application of the principle of strain equivalence. The effective stress $\tilde{\sigma}$ must verify the general form of the law of elasticity,

$$\varepsilon_{ij}^{e} = \frac{1+\nu}{E}\tilde{\sigma}_{ij} - \frac{\nu}{E}\tilde{\sigma}_{kk}\delta_{ij}.$$

The identification of the two expressions for the elastic strain ε_{ij}^{e} gives the effective stress as a function of $\langle\sigma\rangle$, $\langle-\sigma\rangle$, D, h and ν if the law of elasticity is written in the form of the following equivalent expression:

$$\varepsilon_{ij}^{e} = \left\{ \begin{aligned} &\frac{1+\nu}{E}\left[\frac{\langle\sigma_{ij}\rangle + \frac{\nu}{1-2\nu}(\langle\sigma_{kl}\rangle\delta_{kl} - \langle\sigma_{kk}\rangle)\delta_{ij}}{1-D} - \frac{\langle-\sigma_{ij}\rangle + \frac{\nu}{1-2\nu}(\langle-\sigma_{kl}\rangle\delta_{kl} - \langle-\sigma_{kk}\rangle)\delta_{ij}}{1-Dh}\right] \\ &-\frac{\nu}{E}\left(\left[\frac{\langle\sigma_{kl}\rangle + \frac{\nu}{1-2\nu}(\langle\sigma_{rs}\rangle\delta_{rs} - \langle\sigma_{rr}\rangle)\delta_{kl}}{1-D} - \frac{\langle-\sigma_{kl}\rangle + \frac{\nu}{1-2\nu}(\langle-\sigma_{rs}\rangle\delta_{rs} - \langle-\sigma_{rr}\rangle)\delta_{kl}}{1-Dh}\right]\delta_{kl}\right)\delta_{ij} \end{aligned} \right\}$$

Care should be taken with respect to the order of the operations; for example,

$$([\langle\sigma_{rr}\rangle\delta_{kl}]\delta_{kl})\delta_{ij} = \langle\sigma_{rr}\rangle 3\delta_{ij},$$

and also

$$t_r\langle\sigma\rangle \neq \langle t_r(\sigma)\rangle$$

or

$$\langle\sigma_{kl}\rangle\delta_{kl} \neq \langle\sigma_{kk}\rangle,$$

$$\tilde{\sigma}_{ij} = \frac{\langle\sigma_{ij}\rangle + \frac{\nu}{1-2\nu}(\langle\sigma_{kl}\rangle\delta_{kl} - \langle\sigma_{kk}\rangle)\delta_{ij}}{1-D} - \frac{\langle-\sigma_{ij}\rangle + \frac{\nu}{1-2\nu}\langle-\sigma_{kl}\rangle\delta_{kl} - \langle-\sigma_{kk}\rangle\delta_{ij}}{1-Dh}$$

Let us introduce more convenient notation:

$$\sigma_H = \frac{1}{3}\sigma_{kk} \text{ the hydrostatic tress;}$$

$$\sigma_H^+ = \frac{1}{3}\langle\sigma_{kl}\rangle\delta_{kl}, \quad \sigma_H^- = \frac{1}{3}\langle-\sigma_{kl}\rangle\delta_{kl};$$

$$\sigma_{ij}^+ = \langle\sigma_{ij}\rangle + \frac{3\nu}{1-2\nu}(\sigma_H^+ - \langle\sigma_H\rangle)\delta_{ij};$$

$$\sigma_{ij}^- = \langle-\sigma_{ij}\rangle + \frac{3\nu}{1-2\nu}(\sigma_H^- - \langle-\sigma_H\rangle)\delta_{ij};$$

$$\boxed{\tilde{\sigma}_{ij} = \frac{\sigma_{ij}^+}{1-D} - \frac{\sigma_{ij}^-}{1-Dh}}.$$

The two terms with the factor $\frac{3\nu}{1-2\nu}$ are coupling terms which account for shear effects on the stiffness of the material. They disappear in the one-dimensional case;

which we consider here as an exercise:

$$[\sigma] = \begin{bmatrix} \sigma & 0 & 0 \\ 0 & 0 & 0 \\ 0 & 0 & 0 \end{bmatrix}$$

tension σ^+	compression σ^-
$\sigma_H = \frac{1}{3}\sigma^+$	$\sigma_H = \frac{1}{3}\sigma^-$ (negative)
$\sigma_H^+ = \frac{1}{3}\sigma^+$	$\sigma_H^+ = 0$
$\sigma_H^- = 0$	$\sigma_H^- = -\frac{1}{3}\sigma^-$ (positive)
$\sigma_H^+ - \langle\sigma_H\rangle = 0$	$\sigma_H^- - \langle -\sigma_H\rangle = 0$
$\tilde{\sigma}^+ = \frac{\sigma^+}{1-D}$	$\tilde{\sigma}^- = \frac{\sigma^-}{1-Dh}$.

The damage mechanics in unilateral conditions may be developed as it has been done all through this chapter by simply using the new expression for the effective stress in the potentials.

Damage equivalent stress in unilateral conditions

From the state specific enthalpy potential one can also derive the strain energy release rate density Y and the damage equivalent stress σ^*.
Starting from ψ_e^*,

$$\begin{aligned}\psi_e^* &= \frac{1}{2\rho E(1-D)}\left[(1+\nu)\langle\sigma_{ij}\rangle\langle\sigma_{ij}\rangle - \nu\langle\sigma_{kk}\rangle^2\right] \\ &\quad + \frac{1}{2\rho E(1-Dh)}\left[(1+\nu)\langle-\sigma_{ij}\rangle\langle-\sigma_{ij}\rangle - \nu\langle-\sigma_{kk}\rangle^2\right], \\ \overline{Y} &= -\rho\frac{\partial\psi_e^*}{\partial D} = -\frac{1}{2E(1-D)^2}\left[(1+\nu)\langle\sigma_{ij}\rangle\langle\sigma_{ij}\rangle - \nu\langle\sigma_{kk}\rangle^2\right] \\ &\quad - \frac{h}{2E(1-Dh)^2}\left[(1+\nu)\langle-\sigma_{ij}\rangle\langle-\sigma_{ij}\rangle - \nu\langle-\sigma_{kk}\rangle^2\right]\end{aligned}$$

The minus sign comes from the fact that, with damage, the specific enthalpy ψ^* is only a partial Legendre-Fenchel transformation of the specific free energy ψ.

$$Y = \frac{(1+\nu)}{2E}\left[\frac{\langle\sigma_{ij}\rangle\langle\sigma_{ij}\rangle}{(1-D)^2} + \frac{h\langle-\sigma_{ij}\rangle\langle-\sigma_{ij}\rangle}{(1-Dh)^2}\right] - \frac{\nu}{2E}\left[\frac{\langle\sigma_{kk}\rangle^2}{(1-D)^2} + \frac{h\langle-\sigma_{kk}\rangle^2}{(1-Dh)^2}\right].$$

The damage equivalent stress was defined in Section 2.1.2 as the one-dimensional stress (we add here : in tension), which, for the same value of the damage, gives the same value of the strain energy density or, here, of the complementary energy density $w_e^* = \rho\psi_e^*$.
For a tensile stress σ^*,

$$w_e^* = \frac{1}{2E(1-D)}\left[(1+\nu)\sigma^{*2} - \nu\sigma^{*2}\right] = \frac{\sigma^{*2}}{2E(1-D)}.$$

Writing the equality with the expression of w_e^* for the three-dimensional case,

$$\sigma^* = \Big\{(1+\nu)\langle\sigma_{ij}\rangle\langle\sigma_{ij}\rangle - \nu\langle\sigma_{kk}\rangle^2 + \\ + \frac{1-D}{1-Dh}\left[(1+\nu)\langle-\sigma_{ij}\rangle\langle-\sigma_{ij}\rangle - \nu\langle-\sigma_{kk}\rangle^2\right]\Big\}^{1/2}$$

It is easy to check that

for tension (σ^+), $\quad \sigma^* = \sigma^+;$

for compression (σ^-), $\quad \sigma^* = \left(\frac{1-D}{1-Dh}\right)^{1/2} |\sigma^-|.$

2.3.5 Damage Localization and Instability (R. Billardon, I. Doghri, G. Geymonat, 1989)

Damage, as a softening process, may induce localization of strain, as it is the case for plastic shear bands in plasticity. This is the beginning of the process of failure by mesocrack initiation which transforms a volume phenomenon into a surface of separation.

Consider a RVE loaded by a uniform stress $\boldsymbol{\sigma}$ inducing a strain $\boldsymbol{\varepsilon}$ either elastic or plastic, and a damage D. Let $\vec{u}$ be the displacement. Now let us search for the conditions of existence of a surface of discontinuity of the strain or of the strain rate $\dot{\boldsymbol{\varepsilon}}$, as the velocity problem is easier to solve (Figure 2.16).

The constitutive equation of the material is assumed to have been worked out to have the rate form

$$\dot{\sigma}_{ij} = H_{ijkl}\dot{\varepsilon}_{kl},$$

where $\mathbf{H}$ is the tangent stiffness tensor, a function of the state of stress or elastic and plastic strains and of damage as a softening process.

Since this is a problem of continuum mechanics, the displacement and the stress vector are continuous even across the surface $(\vec{n})$. With the notation of the discontinuity of a quantity X across a surface,

$$(X^+ - X^-) = [\![X]\!],$$

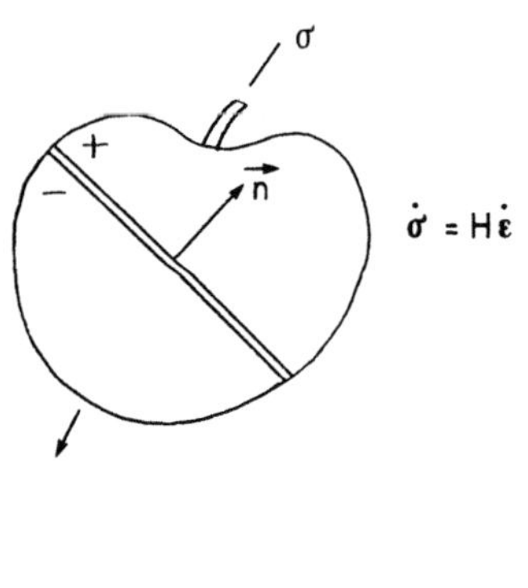

Fig. 2.16. Damage localization in a R.V.E

$$[\![\dot{u}_i]\!] = 0,$$
$$[\![\dot{\sigma}_{ij} n_j]\!] = 0.$$

Using Maxwell's compatibility equation for the general shape of the discontinuity of a second order tensor gradient, $\dot{\boldsymbol{\varepsilon}} = \frac{1}{2}[\operatorname{grad}\dot{\vec{u}} + (\operatorname{grad}\dot{\vec{u}})^T]$, which must be written as

$$[\![\dot{\varepsilon}_{ij}]\!] = \frac{1}{2}(g_i n_j + n_i g_j),$$

where $\vec{g}$ is a vector to be determined:

— $\vec{g} = 0$, no discontinuity;
— $\vec{g} \neq 0$, sufficient condition for the discontinuity to exist.

Let us see for which property this corresponds to **H**, starting with

$$0 = [\![\dot{\sigma}_{ij} n_j]\!] = [\![\dot{\sigma}_{ij}]\!] n_j = [\![H_{ijkl}\dot{\varepsilon}_{kl}]\!] n_j,$$

H being the stiffness tensor of a continuous material, we may assume that it has no discontinuity across the surface $(\vec{n})$: $\mathbf{H}^+ = \mathbf{H}^-$, then:

$$[\![H_{ijkl}\dot{\varepsilon}_{kl}]\!] n_j = H_{ijkl}[\![\dot{\varepsilon}_{kl}]\!] n_j = H_{ijkl}\left[\!\!\left[\frac{1}{2}(g_k n_l + n_k g_l)\right]\!\!\right] n_j = 0.$$

Furthermore, **H** has the minor symmetries,

$$H_{ijkl} = H_{ijlk}, \quad H_{ijkl} = H_{jikl};$$

then,
$$H_{ijkl} g_k n_l = H_{ijkl} n_k g_l$$

and
$$H_{ijkl}\left[\!\!\left[\frac{1}{2}(g_k n_l + n_k g_l)\right]\!\!\right] n_j = n_j H_{ijkl} n_l g_k = 0.$$

$n_j H_{ijkl} n_l$ is a second order tensor $\vec{n}\cdot\mathbf{H}\cdot\vec{n}$. The vector $\vec{n}\cdot\mathbf{H}\cdot\vec{n}\cdot\vec{g}$ has to be zero with a nonzero vector $\vec{g}$ to ensure the existence of the discontinuity of the strain rate $\dot{\boldsymbol{\varepsilon}}$ together with the continuity of the stress vector rate $\dot{\boldsymbol{\sigma}}\cdot\vec{n}$:

$$(n_j H_{ijkl} n_l) g_k = 0, \qquad \text{with } \vec{g} \neq 0.$$

This requires the matrix $[\vec{n}\cdot\mathbf{H}\cdot\vec{n}]$ to be singular; that is,

$$\boxed{\det(\vec{n}\cdot\mathbf{H}\cdot\vec{n}) = 0}.$$

This is a sufficient condition for the existence of a surface of discontinuity of strain, which can be used for damage localization if **H** depends upon D.
If an orientation $\vec{n}$ can be found in an R.V.E. such that $\det(\vec{n}\cdot\mathbf{H}\cdot\vec{n}) = 0$, then a mesocrack will be initiated in the plane perpendicular to $\vec{n}$.
As an exercise, let us work out the simple tension of the plane sheet problem depicted in Figure 2.17:

$$[\sigma] = \begin{bmatrix} \sigma & 0 & 0 \\ 0 & 0 & 0 \\ 0 & 0 & 0 \end{bmatrix}, \quad [\varepsilon] = \begin{bmatrix} \varepsilon & 0 & 0 \\ 0 & -\nu^*\varepsilon & 0 \\ 0 & 0 & -\nu^*\varepsilon \end{bmatrix}.$$

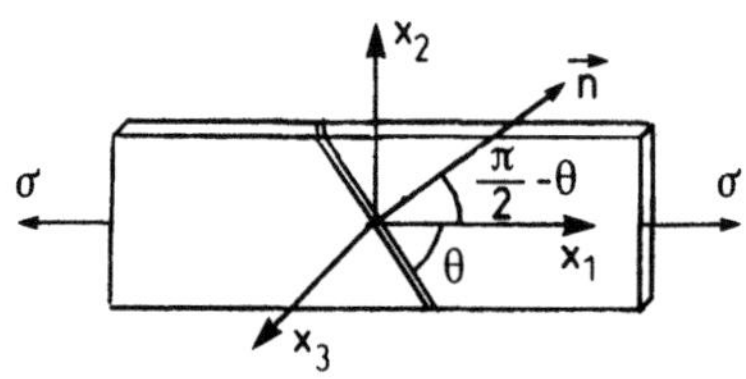

Fig. 2.17. Crack initiation in tension

The constitutive equations of the material do not need to be introduced. Let us only consider a constant contraction coefficient ν^* and the general law

$$\dot{\sigma}_{ij} = H_{ijkl}(\boldsymbol{\varepsilon}, D)\dot{\varepsilon}_{kl}.$$

For the one-dimensional problem:

$$\begin{aligned}\dot{\sigma} &= H_{1111}\dot{\varepsilon} - \nu^*\dot{\varepsilon}(H_{1122} + H_{1133}),\\ 0 &= H_{2211}\dot{\varepsilon} - \nu^*\dot{\varepsilon}(H_{2222} + H_{2233}),\\ 0 &= H_{3311}\dot{\varepsilon} - \nu^*\dot{\varepsilon}(H_{3322} + H_{3333}).\end{aligned}$$

The components of the matrix $[H]$ may be calculated analytically from the constitutive equations of each material.
Nevertheless, this particular case may be solved by considering the discontinuities of the strain rates. The shear strain ε_{12} being zero in the tension problem considered, the discontinuity of its rate is assumed to be zero also. This corresponds to a particular mode, the principal one, among several others also possible. From Maxwell's compatibility equation,

$$[\![\dot{\varepsilon}_{12}]\!] = \frac{1}{2}(g_1 n_2 + n_1 g_2) = 0.$$

For the problem in planes (x_1, x_2) or (x_1, x_3), the components of the unit normal vector $\vec{n}$ to a possible discontinuity line defined by its angle θ with the x_1 axis are

$$\vec{n} : \begin{bmatrix} \sin\theta \\ \cos\theta \\ 0 \end{bmatrix}.$$

$\vec{g}$ is a vector in the (x_1, x_2) plane; let us call it

$$\vec{g} : \begin{bmatrix} \gamma\sin\chi \\ \gamma\cos\chi \\ 0 \end{bmatrix}.$$

Then $[\![\dot{\varepsilon}_{12}]\!] = \dfrac{1}{2}(\gamma\sin\chi\cos\theta + \gamma\sin\theta\cos\chi) = 0$

or $\quad \sin(\chi+\theta) = 0 \rightarrow \chi+\theta = k\pi$ with $k = 0, \pm 1, \pm 2, \ldots,$

which gives the components of $\vec{g}$:

$$\vec{g} : \begin{bmatrix} -\gamma\sin\theta \\ +\gamma\cos\theta \\ 0 \end{bmatrix} \cdot (-1)^k.$$

We may now write the discontinuities of the normal strain rates:

$$[\![\dot{\varepsilon}_{11}]\!] = g_1 n_1 = -\gamma \sin^2\theta \cdot (-1)^k,$$
$$[\![\dot{\varepsilon}_{22}]\!] = g_2 n_2 = +\gamma \cos^2\theta \cdot (-1)^k,$$

and we know from the strain matrix in tension (Figure 2.17) that $\varepsilon_{22} = -\nu^* \varepsilon_{11}$. Assuming this relation is still valid locally,

$$[\![\dot{\varepsilon}_{22}]\!] = -\nu^* [\![\dot{\varepsilon}_{11}]\!].$$

This leads to

$$\cos^2\theta = \nu^* \sin^2\theta \text{ or } \tan\theta = \pm\sqrt{\frac{1}{\nu^*}}.$$

This important result shows that the orientation of the mesocrack that will be initiated does not depend upon the entire constitutive equation, but only upon the contraction coefficient.

Two cases are of practical interest:

* $\nu^* = 0.5$ corresponds to rigid, perfectly plastic damageable materials like most metals close to rupture;

$\theta = \text{Arc}\tan\sqrt{\frac{1}{0.5}} = \pm 54^{\underline{0}}$ is the classic angle of the Luder's bands which may be easily observed by experiments;

* $\nu^* = \nu$, Poisson's ratio, corresponding to elastic damageable materials:
  $\nu = 0.2$ is the value for concrete: $\theta = +65.9^{\underline{0}}$;
  $\nu = 0.3$ is the value for cast iron $\theta = +61.3^{\underline{0}}$;
  $\nu = 0.4$ is the value for plexiglass $\theta = +57.7^{\underline{0}}$.

The stress strain condition for damage localization is deduced from the equilibrium condition, namely, no discontinuity upon the stress vector, together with the constitutive equations:

$$\dot{\sigma}_{ij} = H_{ijkl}\dot{\varepsilon}_{kl},$$
$$[\![\dot{\sigma}_{ij} n_j]\!] = [\dot{\sigma}_{ij}] n_j = 0;$$

expanded for the first component,

$$[\![\dot{\sigma}_{11}]\!] = H_{1111}[\![\dot{\varepsilon}_{11}]\!] + H_{1122}[\![\dot{\varepsilon}_{22}]\!] + H_{1133}[\![\dot{\varepsilon}_{33}]\!] = 0.$$

The directions x_2 and x_3 playing identical roles, we have

$$H_{1122} = H_{1133}.$$

Also $$[\![\dot{\varepsilon}_{22}]\!] = -\nu^*[\![\dot{\varepsilon}_{11}]\!] \text{ and } [\![\dot{\varepsilon}_{33}]\!] = -\nu^*[\![\dot{\varepsilon}_{11}]\!];$$

then $$H_{1111}[\![\dot{\varepsilon}_{11}]\!] - 2\nu^* H_{1122}[\![\dot{\varepsilon}_{11}]\!] = 0;$$

when localization occurs: $$[\![\dot{\varepsilon}_{11}]\!] \neq 0,$$

which implies $$\frac{H_{1111}}{2H_{1122}} = \nu^*.$$

This localization condition corresponds for this tension problem to the peak of the stress strain curve:

$$\dot{\sigma}_{11} = H_{1111}\dot{\varepsilon}_{11} + H_{1122}\dot{\varepsilon}_{22} + H_{1133}\dot{\varepsilon}_{33},$$

as $$H_{1122} = H_{1133} \text{ and } \dot{\varepsilon}_{33} = \dot{\varepsilon}_{22} = -\nu^* \dot{\varepsilon}_{11},$$

$$\dot{\sigma}_{11} = (H_{1111} - 2\nu^* H_{1122})\dot{\varepsilon}_{11} = 0,$$

$$\frac{d\sigma_{11}}{d\varepsilon_{11}} = 0.$$

This result is a necessary and sufficient condition for damage localization; it does not give the condition for crack initiation, which may occur later by the rupture criterion $D = D_c$.

To summarize, the mesocrack:

— is initiated for $D = D_c$;
— growths in the direction normal to $\vec{n}$ defined by $\det(\vec{n} \cdot \mathbf{H} \cdot \vec{n}) = 0$;
— has a size of the order of magnitude of the RVE, also determined by the correlation between damage mechanics and fracture mechanics, $l = \dfrac{G_e}{Y_e}$.

2.3.6 Exercise on the Fiber Bundle System (J. Hult 1982, Chrzanowski 1986)

A nice way to understand the mechanical development of damage is to play with a set of parallel tensile fibers as represented in Figure 2.18(a).

There is no interaction between the fibers but they are all subjected to the same elongation. In fact, the analysis is simplified by considering the continuum model of Figure 2.18(b).

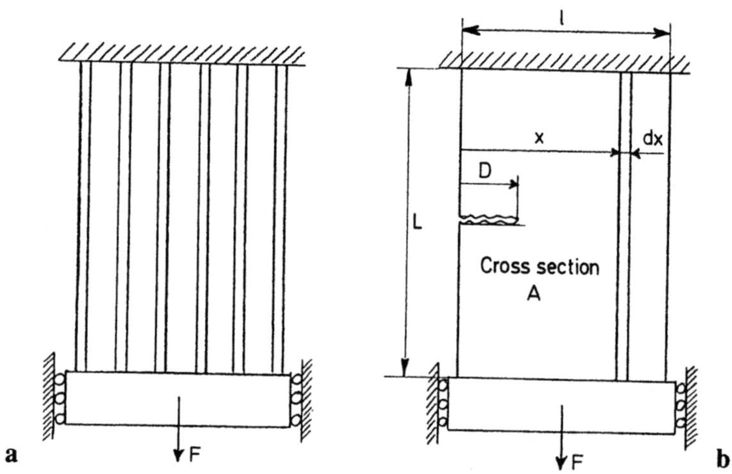

Fig. 2.18. Fiber bundle model

Brittle and ductile failures

The fibers (dx in the continuum model) are linearly elastic with a Young's modulus $E(x)$ and brittle with an ultimate strength $\sigma_R(x)$. They are arranged in such a way that $E(x)$ and $\sigma_R(x)$ are linear monotonically increasing functions of x:

$$E(x) = \overline{E}[1 + a(2x - 1)],$$
$$\sigma_R(x) = \overline{\sigma}_R[1 + b(2x - 1)],$$

where $\overline{E}$ and $\overline{\sigma}_R$ are the values of E and σ_R in the middle of the fiber bundle system ($x = 1/2$). a and b are characteristic coefficients.
If $0 < b < 1$ and $a < b$, *show that for increasing value of the load* F, *the first fiber rupture will occur at* $x = 0$ *for* $F = F_0$:

$$F_0 = \int_0^1 E(x)\varepsilon A\, dx = \overline{\sigma}_R A \frac{1 - b}{1 - a}.$$

As the load increases, consecutive fibers fail, let us call $x = D$ the rupture front defined by

$$\sigma_{(D)} = \varepsilon E_{(D)} = \sigma_{R(D)}.$$

D is the relative surface of decohesion ; it is a measure of the damage of the fiber bundle system.
Determine the following relation between the damage D *and the strain* ε:

from $$\varepsilon E_{(x=D)} = \sigma_{R(x=D)} \rightarrow D = \frac{\overline{E}\varepsilon(1 - a) - \overline{\sigma}_R(1 - b)}{2(\overline{\sigma}_R b - \overline{E}\varepsilon a)}.$$

Determine the relation between the load F *and the damage* D *in correspondence with the three particular cases given in Figure 2.19*:

$$F = \overline{\sigma}_R A \frac{(1 - D + aD - aD^2)(1 - b + 2bD)}{1 - a + 2aD}.$$

The first case is interesting because it shows the possibility of gradual rupture from $F = F_o$ until $\partial F/\partial D = 0$ in a way somewhat similar to ductile damage failure (but here without any plastic strain).
From $\partial F/\partial D = 0$, *deduce the following equation from which the smallest root gives* D_c (This is a good exercise for practicing algebra!):

$$8a^2bD^3 + (2a^2 + 10ab - 12a^2b)D^2 + (2a + 4b - 2a^2 - 10ab + 6a^2b)D +$$
$$+ 1 - 3b + a^2 + 2ab - a^2b = 0.$$

Note that D_c is much smaller than 1.

Creep rupture

The fibers are now considered to be linearly viscoplastic, with the constitutive equation:

$$\dot{\varepsilon}(t) = \frac{\sigma(x, t)}{K_N(x)} \text{ where } K_N = \overline{K}_N[1 + c(2x - 1)].$$

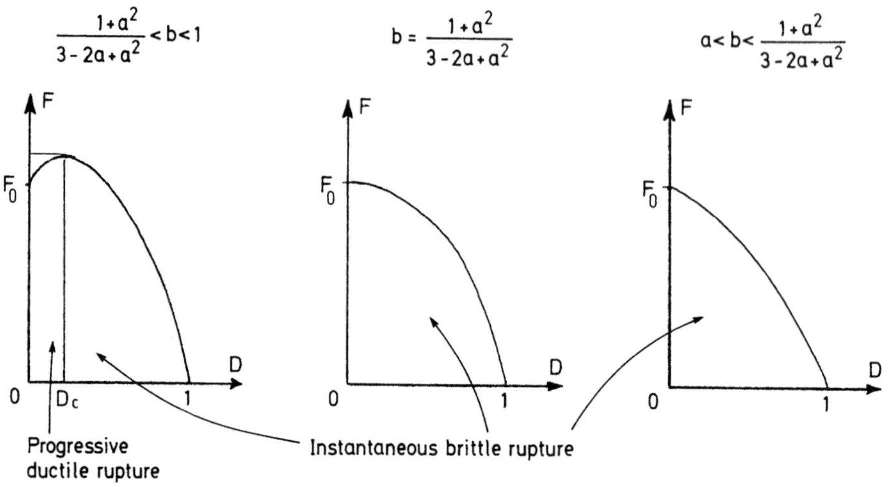

Fig. 2.19. Motion of the rupture damage front D in the fiber bundle model

The elastic strain is neglected but the ultimate tensile stress is still

$$\sigma_R = \overline{\sigma}_R[1 + b(2x - 1)],$$

with $\quad 0 < b < 1, \quad -1 \leq c \leq 1$ and $b > c$.

As the creep strain progresses, the volume remains constant.

Determine the true stress $\frac{F}{A}$ *as a function of the engineering stress* $\frac{F}{A_o}$ *where* A_o *is the initial cross-section area,*

$$d\varepsilon = \frac{dL}{L}; \quad L \cdot A = L_o A_o \rightarrow \frac{F}{A} = \frac{F}{A_o} \exp \varepsilon.$$

The load F is constant ; it induces creep and failure of the fibers. *Determine the creep rate* $\dot{\varepsilon}$ *as a function of time before failure, with the initial condition* $t = 0 \rightarrow \varepsilon = 0$.

From the equilibrium equation

$$\int_0^1 \sigma(x,t)\, dx = \frac{F}{A} \rightarrow \dot{\varepsilon} = \frac{F}{A_o \overline{K}_N} \exp \varepsilon$$

and by integration,
$$\varepsilon = -\mathrm{Ln}\left(1 - \frac{Ft}{A_o \overline{K}_N}\right),$$

it follows that
$$\dot{\varepsilon} = \frac{\frac{F}{A_o}}{\overline{K}_N - \frac{F}{A_o} t}.$$

Determine the time at which the first fiber breaks at $x = 0$.
The rupture condition is

$$\sigma_{(x=0,t_o)} = \sigma_{R(x=0)},$$

$$\overline{K}_N(1-c)\dot{\varepsilon}_{(t_o)} = \overline{\sigma}_R(1-b),$$

$$t_o = \frac{A_o\overline{K}_N}{F}\left[1 - \frac{\frac{F}{A_o}(1-c)}{\overline{\sigma}_R(1-b)}\right].$$

* if
$$\frac{F}{A_o} \geq \overline{\sigma}_R\frac{1-b}{1-c} \rightarrow t_o = 0,$$

there is immediate rupture of the fiber upon load application.

* if
$$\frac{F}{A_o} \leq \overline{\sigma}_R\frac{1-b}{1-c} \rightarrow t_o \text{ is finite,}$$

there is delayed rupture of an incubation time $t_o > 0$. These two cases are represented in Figure 2.20.
The failure propagation is governed by two equations:

* failure at
$$x = D: \quad \sigma_{(t,x=D)} = \sigma_{R(x=D)}$$

or
$$\dot{\varepsilon}_{(t)}\overline{K}_N[1+c(2D-1)] = \overline{\sigma}_R[1+b(2D-1)];$$

* equilibrium
$$\int_D^1 \sigma_{(x,t)}\,dx = \frac{F}{A}$$

or
$$\int_D^1 \dot{\varepsilon}_{(t)}\overline{K}_N[1+c(2x-1)]\,dx = \frac{F}{A_o}\exp\varepsilon_{(t)}.$$

Deduce the following relation between the damage $D(t)$ and the creep strain $\dot{\varepsilon}$:

$$\frac{2(b-c)(1+cD)(1-D)-[1+c(2D-1)]^2[1+b(2D-1)]}{[1+c(2D-1)]^2}\dot{D} = \frac{F}{A_o\overline{\sigma}_R}\dot{\varepsilon}\exp\varepsilon.$$

(This is also a good training in algebra!)
After the first fiber is broken, either an instantaneous break of the whole fiber bundle or gradual failure propagation will ocur, depending upon the relative values of b and c.
Express the damage rate when the first fiber breaks.

$$D = 0\left\{\begin{array}{l} \dot{\varepsilon} = \dfrac{F}{A_o\overline{K}_N}\exp\varepsilon \\ \dot{\varepsilon} = \dfrac{\overline{\sigma}_R}{\overline{K}_N}\dfrac{1-b}{1-c} \end{array}\right\} \quad \dot{D} = \frac{(1-c)(1-b)}{2(b-c)-(1-c)^2((1-b)}\frac{\overline{\sigma}_R}{\overline{K}_N}.$$

Note that the damage rate does not depend upon the load.

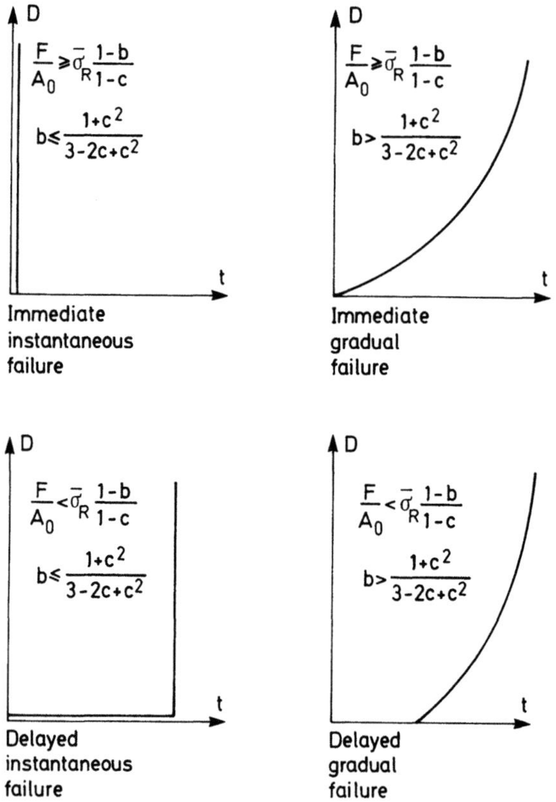

Fig. 2.20. Creep rupture of the fiber bundle model

Determine the conditions of the development of damage:

* instantaneous rupture of the bundle:

$$\dot{D} = \infty;$$

$$\text{if} \quad b \leq \frac{1+c^2}{3-2c+c^2};$$

* gradual development of the damage:

$$\dot{D} \text{ finite,}$$

$$\text{if} \quad b > \frac{1+c^2}{3-2c+c^2}.$$

These two cases appear in Figure 2.20.

Courageous readers can also calculate the time to rupture and the critical value of the damage $x = D_c$ which produces the final instantaneous failure (some approximations must be introduced)!

Chapter 3

Kinetic Laws of Damage Evolution

The physical nature of damage has been defined in the first chapter. The definition of the damage variable as the effective surface density of microcracks in a Representative Volume Element associated with the effective stress concept and the principle of equivalence has given rise to methods of damage measurement through changes in elasticity or plasticity. These concepts have been generalized to the three-dimensional case in the second chapter by means of two potentials. Accordingly, state coupling occurs between elastic strain and damage, and kinetic coupling takes place between plastic strain and damage, which allowing us to calculate strains and damage up to failure if the constitutive equations for the damage are known.

Many models have been proposed in the past; their basic properties are contained in a unified formulation of the potential of dissipation, which will be studied in detail in this chapter. After the kinetic law of isotropic damage is derived, it is applied to several classical cases of loading giving rise to different kinds of damage such as brittle, quasi-brittle, ductile, and low cycle fatigue or high cycle fatigue, with special emphasis on the particular properties of metals, ceramics, polymers, composites and concrete.The common main feature is the proportionality of the damage rate to the strain energy density release rate and to the accumulated plastic strain rate beyond a plastic strain threshold and up to a critical value of the damage variable. The accumulated plastic strain which governs the damage is defined on the meso R.V.E. or at the microscale when the damage is very localized. The case of perfect plasticity is of great interest for further developments.

3.1 Unified Formulation of Damage Laws, (J. Lemaitre 1987)

In Section 2.1.4, the potential of dissipation was written as

$$F = (\tilde{\boldsymbol{\sigma}}^D - \mathbf{X}^D)_{eq} - R - \sigma_y + \frac{3}{4X_\infty} X_{ij}^D X_{ij}^D + F_D(\overline{Y}; (r, D)),$$

with no analytical expression for the damage term F_D from which the damage rate $\dot{D}$ is derived

$$\dot{D} = -\frac{\partial F}{\partial \overline{Y}}\dot{\lambda} = \frac{\partial F_D}{\partial Y}\dot{\lambda},$$

with $$\dot{\lambda} = \dot{r} = \dot{p}(1 - D),$$

and
$$\dot{p} = \left(\frac{2}{3}\dot{\varepsilon}^p_{ij}\dot{\varepsilon}^p_{ij}\right)^{1/2}.$$

The choice of this function F_D is of course the key to representing the damage evolution. It is a matter of logic through the "State Kinetic Coupling theory", of physical observations, of experimental data and of micromechanical mechanism modeling to quantify the effects. Like Hooke's law for elasticity or the Prandtl-Reuss law for plasticity, a unique kinetic law of damage evolution may be developed to model the general trends of all kinds of damage in many kinds of materials.

3.1.1 General Properties and Formulation

Let us list the main trends and properties of damage as presented in Chapter 1.

— Damage is always related to some irreversible strain either at the microlevel or the mesolevel. This property is taken into account in the damage law by the multiplier $\dot{\lambda}$, which is proportional to the accumulated plastic strain. When damage occurs,
$$\dot{D} = \frac{\partial F_D}{\partial Y}\dot{p}(1-D).$$
The variable $\dot{p}$, which governs the damage evolution, also gives the irreversible nature of the damage, as $\dot{p}$ is always positive or zero.

— As the accumulated plastic strain increases from zero, the damage remains equal to zero during the nucleation of microcracks. This corresponds to the accumulation of micro-stresses, or dislocations in metals, generating microcracks. A one-dimensional damage threshold related to the plastic strain ε_{pD} has already been introduced in Sections 1.2.4 and 2.1.4. As the equation of damage is governed by the accumulated plastic strain, and as $p = \varepsilon_p$ in one-dimensional monotonic loading, it is logical to introduce a threshold on the variable p, or on r, as $r = p$ when $D = 0$:
$$\dot{D} = \frac{\partial F_D}{\partial Y}\dot{p}(1-D) \text{ if } p \geq p_D;$$
$$\dot{D} = 0 \text{ if } p < p_D.$$
This allows us to introduce a step function in the potential F_D which has to be a function of the state variable r taken as a parameter:
$$H(r-p_D) = \begin{cases} 1 \text{ if } r \geq p_D \\ 0 \text{ if } r < p_D \end{cases}$$
In monotonic uniaxial loading, p_D may be identified with the uniaxial damage threshold ε_{pD} but in fatigue or creep processes, p_D is a function of the applied stress as will be established in Section 3.1.2.

— On the basis of a thermodynamical analysis, the main causal variable for the damage is the strain energy density release rate Y, as it is the dual variable of the flux $\dot{D}$. Then F_D must be a function of Y:

$$F_D = F_D(Y \ldots).$$

— Another important feature of fracture is the influence of the triaxiality ratio $\frac{\sigma_H}{\sigma eq}$, ($\sigma_H$ is the hydrostatic stress, σ_{eq} is the von Mises equivalent stress). The modeling of this effect is contained in the expression of Y by the triaxiality factor R_ν (see Section 2.1.2).

$$Y = \frac{\sigma_{eq}^2 R_\nu}{2E(1-D)^2}, \quad R_\nu = \frac{2}{3}(1+\nu) + 3(1-2\nu)\left(\frac{\sigma_H}{\sigma_{eq}}\right)^2.$$

— In order to choose the proper and simplest expression for F_D let us recall the kinetic damage relations obtained by micromechanics for particular mechanisms in Section 2.3.

  * Brittle damage by fatigue growth of microcracks:

$$\dot{D} = \frac{\eta c E^{\eta/2} d^{\eta/2} e}{l^2 n^{\eta/2-1}} Y^{\eta/2-1} \dot{Y},$$

η being of the order of 4 for most materials, D is proportional to $Y \cdot \dot{Y}$:

$$\dot{D} \sim Y \cdot \dot{Y}.$$

Here, no plasticity has been introduced, but it always exists at the microscale at the crack tips of the microcracks and it is possible at least formally to relate $\dot{Y}$ to the accumulated microplastic strain rate $\dot{p}^\mu$ through a plasticity constitutive equation:

$$Y = \frac{\sigma_{eq}^2 R\nu}{2E(1-D)^2} \rightarrow \dot{Y}_{(\dot{\sigma}_{eq})}, \quad \dot{\sigma}_{eq(\dot{p}^\mu)} \leftarrow \sigma_{eq(p^\mu)}.$$

  * Ductile damage by nucleation of microcavities:

$$\dot{D} = \frac{l}{d} E_T \left(A + B\frac{\sigma_H}{\sigma_{eq}}\right) \dot{p}.$$

  * Ductile damage by enlargement of microcavities:

$$\dot{D} = 0.57D \left[\exp\left(\frac{3}{2}\frac{\sigma_H}{\sigma_{eq}}\right)\right] \dot{p}.$$

A general qualitative statement of these three results can be obtained by considering $\dot{D}$ to be proportional to Y (function of $\frac{\sigma_H}{\sigma_{eq}}$ and $\dot{p}$):

$$\dot{D} \sim Y\dot{p}$$

or

$$F_D \sim Y^2.$$

— Here, as in every constitutive equation, a scale factor like $\frac{\eta c E^{\eta/2} d^{\eta/2} e}{l^2 n^{\eta/2-1}}$, $\frac{l}{d} E_T$ or 0.57 seen above must be introduced. Let us take

$$S(1-D),$$

where S is a material constant. The term $(1-D)$ is considered here to be cancelled, with $(1-D)$ coming from $\dot{\lambda} = \dot{p}(1-D)$, because experiments show a non decreasing damage rate when Y and $\dot{p}$ are constant:

$$F_D \sim \frac{Y^2}{S(1-D)}.$$

— The rupture criterion $D = D_c$ introduced in one dimension in Section 1.2.4, which reduces to $D = D_{1c}$ in the pure tension reference case, also defines conditions for mesocrack initiation in three dimensions. D_c will be established as a function of D_{1c}, σ_u and the loading $\tilde{\sigma}^*$ in Section 3.1.3.

— Finally, according to the quantitative properties listed above, the damage potential is logically written as:

$$F_D(Y;(r,D)) = \frac{Y^2}{2S(1-D)} H_{(r-p_D)}.$$

The factor 1/2 is used here to avoid 2 in the derivation:

$$\dot{D} = \frac{\partial F_D}{\partial Y}\dot{\lambda} = \frac{Y}{S(1-D)}\dot{p}(1-D)H_{(r-p_D)}$$

or, with $r = p$ when $D = 0$,

$$\boxed{\dot{D} = \frac{Y}{S}\dot{p}H_{(p-p_D)}},$$

with the rupture condition for crack initiation,

$$D = D_c.$$

— Three material parameters are introduced to characterize the damage evolution:
  * S → the energy strength of damage;
  * p_D → the damage threshold function of the material and the loading (see Section 3.1.2.)
  * D_c → the critical damage, a function of the material and the loading (see Section 3.1.3).

The effects of the temperature T are taken into account by the variation of these coefficients with T and by the accumulated plastic strain rate $\dot{p}$, also a function of the temperature.

Several properties not directly introduced in the formulation will be observed as consequences:

— the nonlinear accumulation of damage;
— the effect of mean stress in fatigue;
— the nonlinear interaction of different kinds of damage;
— the brittleness of materials as the applied stress or the triaxiality increase.

3.1.2 Stored Energy Damage Threshold

The damage threshold p_D, or ε_{pD} in one dimension, corresponds to a nucleation of microcracks which does not produce any change in the mechanical properties.

It is related to the amount of energy which is stored in the material. Experiments in fatigue have shown that the total plastic strain energy dissipated may reach tremendous values before failure but the stored energy remains constant at micro-crack initiation, no matter what the stress is.
This stored energy is the result of microstress concentrations which develop in the neighborhood of dislocation networks in metals and of inhomogeneities in other materials. For a unit volume, it is equal to the difference between the total plastic strain energy $\int_0^t \sigma_{ij}\dot{\varepsilon}^p_{ij}dt$ and the energy dissipated in heat given by the Clausius-Duhem inequality of the second principle of thermodynamics (see Section 2.1.3). For an isothermal transformation ($\vec{q} = 0$) of a material having kinematic and isotropic strain hardenings and no damage ($\dot{D} = 0 \rightarrow \dot{r} = \dot{p}$), the power dissipated is.

$$\phi = \sigma_{ij}\dot{\varepsilon}^p_{ij} - R\dot{p} - X^D_{ij}\dot{\alpha}_{ij} \geq 0,$$

which may be calculated from:

— the potential of dissipation (Section 2.1.4),

$$F = (\boldsymbol{\sigma}^D - \mathbf{X}^D)_{eq} - R - \sigma_y + \frac{3}{4X_\infty}X^D_{ij}X^D_{ij};$$

— its associated normality flow rule,

$$\dot{\varepsilon}^p_{ij} = \frac{\partial F}{\partial \sigma_{ij}}\dot{\lambda}, \quad \dot{p} = -\frac{\partial F}{\partial R}\dot{\lambda} = \dot{\lambda}, \quad \dot{\alpha}_{ij} = -\frac{\partial F}{\partial X_{ij}}\dot{\lambda};$$

— and the yield criterion (Section 2.1.3)

$$f = (\boldsymbol{\sigma}^D - \mathbf{X}^D)_{eq} - R - \sigma_y = 0,$$

$$\phi = \left(\sigma_{ij}\frac{\partial F}{\partial \sigma_{ij}} + R\frac{\partial F}{\partial R} + X^D_{ij}\frac{\partial F}{\partial X^D_{ij}}\right)\dot{p} \geq 0,$$

$$\phi = \left[\sigma_{ij}\frac{3}{2}\frac{\sigma^D_{ij} - X^D_{ij}}{(\boldsymbol{\sigma}^D - \mathbf{X}^D)_{eq}} - R + X^D_{ij}\left(-\frac{3}{2}\frac{\sigma^D_{ij} - X^D_{ij}}{(\boldsymbol{\sigma}^D - \mathbf{X}^D)_{eq}} + \frac{3}{2X_\infty}X^D_{ij}\right)\right]\dot{p},$$

$$\phi = \left[(\boldsymbol{\sigma}^D - \mathbf{X}^D)_{eq} - R + \frac{3}{2X_\infty}X^D_{ij}X^D_{ij}\right]\dot{p},$$

$$\phi = \left(\sigma_y + \frac{3}{2X_\infty}X^D_{ij}X^D_{ij}\right)\dot{p}.$$

Then, the stored energy w_s as a function of time is

$$w_{s(t)} = \int_0^t \sigma_{ij}\dot{\varepsilon}^p_{ij}dt - \int_0^t \left(\sigma_y + \frac{3}{2X_\infty}X^D_{ij}X^D_{ij}\right)\dot{p}dt.$$

Some approximations make this formula much easier to deal with:
Neglecting the effect of the kinematic hardening,

$$X^D_{ij} = 0 \rightarrow \dot{\varepsilon}^p_{ij} = \frac{\partial F}{\partial \sigma_{ij}}\dot{\lambda} = \frac{3}{2}\frac{\sigma^D_{ij}}{\sigma_{eq}}\dot{p},$$

$$w_s = \int_0^t \frac{3}{2} \frac{\sigma_{ij}^D \sigma_{ij}^D}{\sigma_{eq}} \dot{p} dt - \int_0^t \sigma_y \dot{p} dt$$

or with,

$$\sigma_{eq} = \left(\frac{3}{2} \sigma_{ij}^D \sigma_{ij}^D \right)^{1/2},$$

$$w_s = \int_0^p (\sigma_{eq} - \sigma_y) dp.$$

This formula may be used; however, it is still easier to assume a perfectly plastic material of plastic threshold $\sigma_{eq} \geq \sigma_y$, whose stored energy is a function of the difference between σ_{eq} and the fatigue limit σ_f, that is the stress, still in the conventional elastic range below which no damage may occur. It is identified as the stress amplitude which corresponds to a very large number of cycles to failure in a tension-compression fatigue test: $N_R \simeq 10^6$ to 10^7 cycles. Some values are given in the chart in Section 3.5.
Then, as σ_{eq} = const.

$$w_s = (\sigma_{eq} - \sigma_f) p.$$

The characteristic value of that energy corresponding to microcrack initiation ($D = 0$, $\dot{D} > 0$) is taken as that of the pure tension reference case having a damage threshold ε_{PD} and a plastic threshold σ_u, the ultimate stress.

$$(\sigma_u - \sigma_f) \varepsilon_{PD}.$$

Writing the equality of these two energies gives the value of the damage threshold p_D for any kind of loading:

$$(\sigma_{eq} - \sigma_f) p_D = (\sigma_u - \sigma_f) \varepsilon_{PD},$$

$$\boxed{p_D = \varepsilon_{PD} \frac{\sigma_u - \sigma_f}{\sigma_{eq} - \sigma_f}}.$$

Remember that in this formula the material is considered to be perfectly plastic. Then for a varying loading $\sigma_{eq}(t)$, a more accurate calculation consists in performing the integration for w_s!

3.1.3 Three-Dimensional Rupture Criterion

A one-dimensional rupture criterion was introduced in Section 1.2.4 as a relation between the applied stress at rupture and a critical value of the damage depending upon the material and the loading:

$$\frac{\sigma}{1 - D_c} = \sigma_\infty.$$

Applied to the pure tension test, which is taken as a reference, $D_c = D_{1c}$, $\sigma = \sigma_R$, the stress to rupture, and $\sigma_\infty \simeq \sigma_u$, the ultimate stress (whose values for several materials may be found in the table in Section 3.5). We then obtain

$$\frac{\sigma_R}{1-D_{1c}} = \sigma_u.$$

In three dimensions, the damage criterion introduced in Section 2.1.2 allows us to generalize this one-dimensional relation of rupture.
Let us study the condition of stability of the damage process related to a non positive evolution of the density of kinetic energy k:

$$\dot{k} \leq 0.$$

The first principle of thermodynamics, which gives the balance of energy, is written as:

$$\dot{e}+\dot{k} = p_x + Q - Y_c\dot{D}.$$

The additional term $-Y_c\dot{D}$ is the power disspated to create new micro free surfaces acting as variable boundaries; it is characterized for each material by a critical energy Y_c.
$\dot{e}$ is the internal energy density rate. Assuming the stage of instability to be brittle, $\dot{e}$ is the sum of the elastic strain power density $\dot{w}_e$ and the heat rate Q received by the RVE:

$$\dot{e} = \dot{w}_e + Q.$$

p_x is the external force power density equal to the internal force power density if the inertia forces are neglected for the condition $\dot{k}$ approaching zero:

$$p_x = \sigma_{ij}\dot{\varepsilon}_{ij} = \sigma_{ij}\dot{\varepsilon}^e_{ij}$$

since only elastic energy is involved in brittle processes.
The equation for the first principle of thermodynamics becomes:

$$\dot{k} = \sigma_{ij}\dot{\varepsilon}^e_{ij} - \dot{w}_e - Y_c\dot{D} \leq 0.$$

From the elasticity potential (see Section 2.1.1),

$$w_e = \frac{1}{2}a_{ijkl}\varepsilon^e_{ij}\varepsilon^e_{kl}(1-D)$$

$$\dot{w}_e = a_{ijkl}\varepsilon^e_{ij}\dot{\varepsilon}^e_{kl}(1-D) - \frac{1}{2}a_{ijkl}\varepsilon^e_{ij}\varepsilon^e_{kl}\dot{D}$$

or, with the law of elasticity:

$$\sigma_{kl} = a_{ijkl}\varepsilon^e_{ij}(1-D),$$

and, remembering the definition of the strain energy density release rate (Section 2.1.1),

$$Y = \frac{1}{2}a_{ijkl}\varepsilon^e_{ij}\varepsilon^e_{kl},$$

$$\dot{k} = \sigma_{ij}\dot{\varepsilon}^e_{ij} - \sigma_{kl}\dot{\varepsilon}^e_{kl} + Y\dot{D} - Y_c\dot{D} \leq 0.$$

$\dot{D}$ always being positive, this yields the stability condition

$$Y - Y_c \leq 0$$

and the instability criterion

$$Y > Y_c.$$

Physically, this means that an instability may occur if the energy released by loss of stiffness becomes equal to or greater than the energy needed for the creation of new damage surfaces of decohesions.
In fact, during this stage of instability prior to failure, the strain energy density release rate is almost constant. For example, if a perfectly plastic material of plastic threshold σ_s is submitted to a proportional loading for which R_ν = const. (see Section 2.2.4),

$$\frac{\sigma_{eq}}{1-D} - \sigma_s = 0;$$

$$Y = \frac{\sigma_{eq}^2 R_\nu}{2E(1-D)^2} = \frac{\sigma_s^2 R_\nu}{2E} = \text{const}.$$

The instability may occur at any time during the damage growth when some perturbation slightly increases the quantity Y. Then this instability condition cannot simultaneously be the rupture criterion. From both the practical and theoretical points of view, a physical condition of rupture must be added.
To this end, let us assume that the final stage of damage rupture by atomic decohesion, (that is mesocrack initiation) is governed by the amount of energy dissipated in damage growth:

$$\int_0^{D_c} Y dD = \text{const. at failure.}$$

This quantity may be identified from the uniaxial case in tension, still taken as a reference:

$$\left.\begin{aligned} \sigma_{eq} &= \sigma_R \\ R_\nu &= 1 \\ D &= D_{1c} \end{aligned}\right\} Y = \frac{\sigma_R^2}{2E(1-D_{1c})^2} = \frac{\sigma_u^2}{2E}$$

and

$$\int_0^{D_{1c}} Y dD = \frac{\sigma_u^2}{2E} D_{1c}.$$

In three dimensions, the case of perfect plasticity in proportional loading is, for simplicity, considered as an approximation:

$$\int_0^{D_c} Y dD = \int_0^{D_c} \frac{\sigma_{eq}^2 R_\nu}{2E(1-D)^2} dD = \int_0^{D_c} \frac{\sigma_s^2 R_\nu}{2E} dD = \frac{\sigma_s^2 R_\nu}{2E} D_c = \frac{\tilde{\sigma}^{*2}}{2E} D_c,$$

as $$\tilde{\sigma}^* = \frac{\sigma^*}{1-D}, \quad \sigma^* = \sigma_{eq} R_\nu^{1/2} \text{ and } \sigma_{eq} = \sigma_s(1-D).$$

Then the rupture criterion gives the critical value of the damage as a function of its value for the one-dimensional reference case D_{1c}, the effective damage equivalent stress $\tilde{\sigma}^*$ and the ultimate stress σ_u with the limiting value $D_c = 1$:

$$\frac{\tilde{\sigma}^{*2}}{2E} D_c = \frac{\sigma_u^2}{2E} D_{1c},$$

$$\boxed{D_c = D_{1c} \frac{\sigma_u^2}{\tilde{\sigma}^{*2}} \leq 1}.$$

This formula shows that the critical value of the damage for mesocrack initiation decreases as the effective damage equivalent stress $\tilde{\sigma}^*$ increases either by the stress σ_{eq}, the damage D or by the triaxiality ratio σ_H/σ_{eq} contained in the R_ν expression.

D_{1c} is the critical value of the damage of the case of reference under tension. Any other one-dimensional case (like fatigue, for example) may have a different value of D_c, depending upon the applied effective stress $\tilde{\sigma}$.

$$D_c = D_{1c}\frac{\sigma_u^2}{\tilde{\sigma}^2}.$$

It is interesting to study the different possible cases in the graph $(D_c, \tilde{\sigma}^*)$ of Figure 3.1, where the rupture relation $D_c = D_{1c}\frac{\sigma_u^2}{\tilde{\sigma}^{*2}}$ corresponds to a line which defines:

— a safe domain below which a stress-damage path does not produce crack initiation,

— a rupture domain upon which the material is certainly broken.

One-dimensional examples (Figure 3.2)

— A fatigue case at constant amplitude of stress is represented by path 1: the critical value of the damage decreases as the stress amplitude increases.

— A fatigue case at constant amplitude of strain corresponds to a quasi-constant amplitude of effective stress as shown by path 2.

— A ductile case in tension is the case of reference represented by path 3 for which the critical value of the damage is of course D_{1c}.

— A creep path at constant stress is represented by path 4. The critical value of the damage also decreases as the applied stress increases.

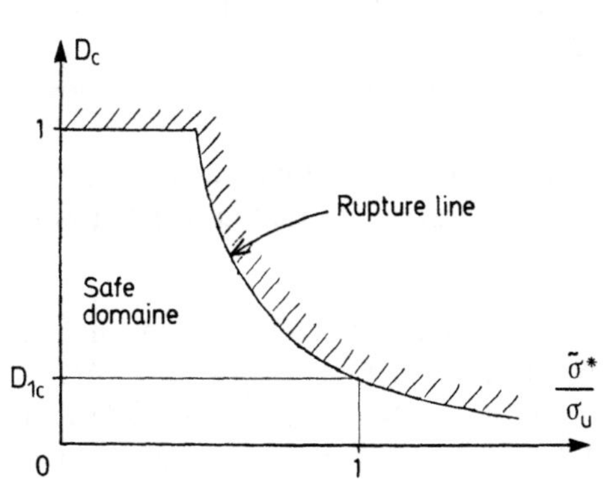

Fig. 3.1. Rupture criterion at the meso-scale

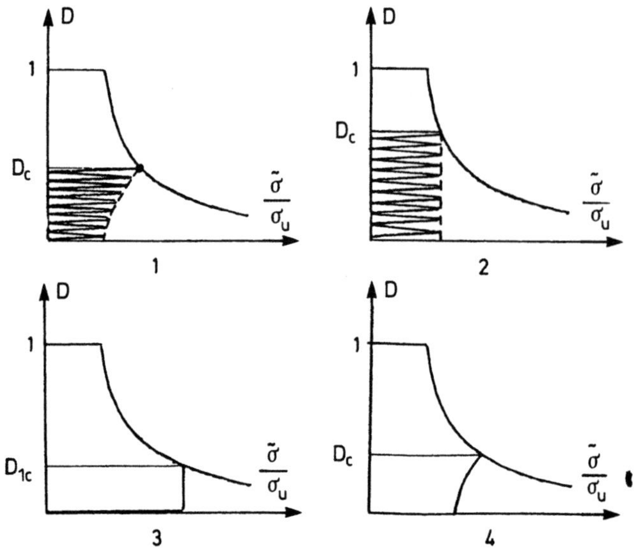

Fig. 3.2. Damage rupture paths. 1) Fatigue $\Delta\sigma$ = const; 2) Fatigue $\Delta\varepsilon$ = const; 3) Ductile case; 4) creep σ = const.

This mesocrack initiation is completed by the localization criterion giving the orientation of the crack developed in Section 2.3.5.
The entire set of equations governing damage evolution can be summarized as follows:

$$\dot{D} = \begin{cases} \dfrac{Y}{S}\dot{p} \text{ if } p \geq p_D \text{ with } p_D = \varepsilon_{PD}\dfrac{\sigma_u - \sigma_f}{\sigma_{eq} - \sigma_f} \\ 0 \text{ if } p < p_D \end{cases}$$

$$\text{Crack initiation if } D = D_c \text{ with } D_c = D_{1c}\frac{\sigma_u^2}{\tilde{\sigma}^{*2}} \leq 1$$

The material parameters are:

— S, ε_{PD} and D_{1c}, which must be determined from damage measurements;
— σ_u and σ_f, which are classical characteristics "easy" to find in handbooks or to identify by tensile and fatigue tests.

In most cases Y is the strain energy density release rate defined in Section 2.1.1. In order to take into account the micro-crack closure effect responsible for different damage behaviors in tension and compression, the only change to be made is in the definition of Y derived in Section 2.3.4 as a function of the crack closure parameter h.

3.1.4 Case of Elastic-Perfectly Plastic and Damageable Materials

In Section 3.1.1 we have observed that damage occurs only if the accumulated plastic strain p (or the plastic strain in tension) has reached a certain threshold p_D:

$$p < p_D \rightarrow \dot{D} = 0 \text{ or } \varepsilon_p < \varepsilon_{PD} \rightarrow \dot{D} = 0.$$

For many materials, this threshold is large enough to assume that the strain hardening is close to saturation making the behavior of the materials perfectly plastic for further straining. This hypothesis corresponds to the saturation of kinematic and the isotropic hardenings.

$$p > p_D \rightarrow \begin{cases} \dot{X}_{ij}^D = 0 \\ \dot{R} = 0 \end{cases}$$

The plasticity criterion is simply

$$f = \tilde{\sigma}_{eq} - \sigma_s = 0,$$

where σ_s is the threshold into consideration for each application. The choice of its value for a given material is somewhat subjective because real materials are never perfectly plastic. But it has to be bounded by σ_y and σ_u:

$$\sigma_y \leq \sigma_s \leq \sigma_u.$$

Taking $\sigma_s = \sigma_u$, the maximum value of the stress on a stress strain curve in tension, where $\dfrac{d\sigma}{d\varepsilon} = 0$, is often a good choice (Figure 3.3 for example).
Let us write the corresponding kinetic laws for plastic strain and damage taken from sections 2.1.4 and 1.1 of the present chapter.

$$\dot{\varepsilon}_{ij}^p = \frac{\partial f}{\partial \sigma_{ij}} \dot{\lambda} \quad \text{if} \begin{cases} f = \dfrac{\sigma_{eq}}{1-D} - \sigma_s = 0 \\ \dot{f} = 0 \text{ with } \dot{\lambda} = \dot{p}(1-D), \end{cases}$$

$$\dot{D} = \frac{Y}{S}\dot{p} \quad \text{if } p > p_D \text{ with } Y = \frac{\sigma_{eq}^2 R\nu}{2E(1-D)^2};$$

from the plasticity criterion:

$$\frac{\partial f}{\partial \sigma_{ij}} = \frac{3}{2}\frac{\sigma_{ij}^D}{\sigma_{eq}}\frac{1}{1-D},$$

$$Y = \frac{\sigma_s^2}{2E} R_\nu;$$

from the consistancy condition:

$$\dot{f} = \frac{\dot{\sigma}_{eq}}{1-D} + \frac{\sigma_{eq}}{(1-D)^2}\dot{D} = 0,$$

$$\dot{\sigma}_{eq} + \sigma_s \dot{D} = 0;$$

then:

$$\dot{D} = \frac{\sigma_s^2}{2ES} R_\nu \dot{p},$$

$$\dot{\varepsilon}_{ij}^p = \frac{3}{2}\frac{\sigma_{ij}^D}{\sigma_s}\frac{\dot{p}}{1-D};$$

or, replacing $\dot{p}$ by its expression in terms of $\dot{D}$ for $p > p_D$

$$\dot{\varepsilon}_{ij}^{p} = \frac{3ES}{\sigma_s^3} \frac{\sigma_{ij}^D}{1-D} \frac{\dot{D}}{R_\nu}.$$

This shows that for perfectly plastic behavior when $p > p_D$, the plastic strain is directly related to the damage.

Another interesting result coming from the consistency condition is

$$\dot{D} = -\frac{\dot{\sigma}_{eq}}{\sigma_s}.$$

Notice that the equivalent stress rate is negative, owing to the softening damage process.

The last interesting result comes from the plasticity criterion written in terms of the elastic strain: the law of elasticity coupled to the damage written in terms of the deviatoric and the hydrostatic stress and strain is

$$\begin{aligned} \varepsilon_{ij}^{e} &= \varepsilon_{ij}^{eD} + \varepsilon_H^e \delta_{ij}, & \varepsilon_H^e &= \frac{1}{3}\varepsilon_{kk}^e, \\ \sigma_{ij} &= \sigma_{ij}^D + \sigma_H \delta_{ij}, & \sigma_H &= \frac{1}{3}\sigma_{kk}, \\ \varepsilon_{ij}^{eD} &= \frac{1+\nu}{E(1-D)}\sigma_{ij}^D, & \varepsilon_H^e &= \frac{1-2\nu}{E(1-D)}\sigma_H. \end{aligned}$$

It follows that

$$\sigma_{eq} = \left(\frac{3}{3}\sigma_{ij}^D \sigma_{ij}^D\right)^{1/2} = \frac{E(1-D)}{1+\nu}\left(\frac{3}{2}\varepsilon_{ij}^{eD}\varepsilon_{ij}^{eD}\right)^{1/2}$$

and

$$f = \frac{\sigma_{eq}}{1-D} - \sigma_s = \frac{E}{1+\nu}\left(\frac{3}{2}\varepsilon_{ij}^{eD}\varepsilon_{ij}^{eD}\right)^{1/2} - \sigma_s$$

or, with $f = 0$,

$$(\varepsilon_{ij}^{eD}\varepsilon_{ij}^{eD})^{1/2} = \sqrt{\frac{2}{3}}\left(\frac{1+\nu}{E}\right)\sigma_s = \text{const}.,$$

which makes the second invariant of the elastic strain deviator constant for any state of stress verifying the plasticity criterion.

In tension, where

$$\begin{aligned} \frac{\sigma_{eq}}{1-D} &= \frac{\sigma}{1-D} = E\varepsilon_e, \\ E\varepsilon_e - \sigma_s &= 0, \\ \varepsilon_e &= \frac{\sigma_s}{E} = \text{const}. \end{aligned}$$

This property is shown schematically in Figure 3.3 for a classical tension test and for a cyclic tension-compression test, strain controlled at constant strain amplitude. It proves that the measure of the damage by the variation of the elasticity modulus and by the stress amplitude drop are equivalent.

The complete set of constitutive equations is

$$\boxed{\begin{aligned} \varepsilon_{ij} &= \varepsilon^e_{ij} + \varepsilon^p_{ij}, \\ \varepsilon^e_{ij} &= \frac{1+\nu}{E}\tilde{\sigma}_{ij} - \frac{\nu}{E}\tilde{\sigma}_{kk}\delta_{ij}, \\ \dot{\varepsilon}^p_{ij} &= \frac{3}{2}\frac{\tilde{\sigma}^D_{ij}}{\sigma_s}\dot{p} \text{ if } \tilde{\sigma}_{eq} - \sigma_s = 0, \\ \dot{D} &= \frac{\sigma_s^2}{2ES}R_\nu \dot{p} \text{ if } p \geq p_D. \end{aligned}}$$

where R_ν may be expressed as a function of the strain:

$\dfrac{\sigma_H}{\sigma_{eq}} = \dfrac{\dfrac{E(1-D)}{1-2\nu}\varepsilon^e_H}{\sigma_s(1-D)}$ with $\varepsilon^e_H = \dfrac{1}{3}t_r(\boldsymbol{\varepsilon} - \boldsymbol{\varepsilon}^p) = \dfrac{1}{3}t_r(\boldsymbol{\varepsilon}) = \varepsilon_H$ since $\boldsymbol{\varepsilon}^p$ is a deviator;

$$\boxed{R_\nu = \frac{2}{3}(1+\nu) + \frac{3}{1-2\nu}\left(\frac{E\varepsilon_H}{\sigma_s}\right)^2}.$$

In order to take into account the micro crack closure effect explained in Section 2.3.4, the assumption of perfect unilateral conditions may be made; it corresponds to a zero value of the crack closure parameter:

$$h = 0.$$

The strain energy density release rate from Section 2.3.4 reduces to

$$Y = \frac{1+\nu}{2E}\frac{<\sigma_{ij}><\sigma_{ij}>}{(1-D)^2} - \frac{\nu}{2E}\frac{<\sigma_{kk}>^2}{(1-D)^2}$$

and, with this expression, the kinetic law of damage remains unchanged:

$$\dot{D} = \frac{Y}{S}\dot{p} \quad \text{if } p \geq p_D$$

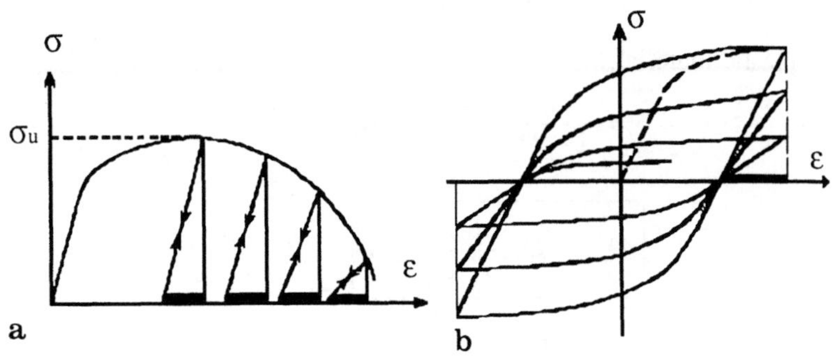

Fig. 3.3. Constant elastic strain. **a** tension test; **b** cyclic tension compression test

Viscoplasticity

The case of nonhardening behavior reduces the set of elastic perfect viscoplasticity constitutive equations from those of Section 2.1.4 to

$$\varepsilon_{ij} = \varepsilon^e_{ij} + \varepsilon^p_{ij},$$

$$\varepsilon^e_{ij} = \frac{1+\nu}{E}\tilde{\sigma}_{ij} - \frac{\nu}{E}\tilde{\sigma}_{kk}\delta_{ij},$$

$$\dot{\varepsilon}^p_{ij} = \frac{3}{2}\frac{\sigma^D_{ij}}{\sigma_{eq}} \operatorname{Ln}\left(1 - \frac{\tilde{\sigma}_{eq} - \sigma_s}{K_\infty}\right)^{-n} \text{ if } \tilde{\sigma}_{eq} \geq \sigma_s,$$

$$\dot{D} = \frac{\sigma^2_{eq} R_\nu}{2ES(1-D)^2}\dot{p} \text{ if } p \geq p_D.$$

with
$$R_\nu = \frac{2}{3}(1+\nu) + \frac{3}{1-2\nu}\left(\frac{E\varepsilon_H}{\tilde{\sigma}_{eq}}\right)^2.$$

Damage threshold

$$p_D = \varepsilon_{PD}\frac{\sigma_u - \sigma_f}{\sigma_{eq} - \sigma_f}.$$

For perfectly plastic materials this becomes

$$\boxed{p_D = \varepsilon_{PD}\frac{\sigma_u - \sigma_f}{\sigma_s - \sigma_f}}$$

and

$$p_D = \varepsilon_{PD} \text{ if } \sigma_s = \sigma_u.$$

Rupture criterion

$$D_c = D_{1c}\frac{\sigma^2_u}{\tilde{\sigma}^{*2}}.$$

For a perfectly plastic material,

$$\tilde{\sigma}^* = \frac{\sigma_{eq}}{1-D}R^{1/2}_\nu = \sigma_s R^{1/2}_\nu.$$

Then

$$\boxed{D_c = D_{1c}\frac{\sigma^2_u}{\sigma^2_s R_\nu} \leq 1}$$

and

$$D_c = \frac{D_{1c}}{R_\nu} \text{ if } \sigma_s = \sigma_u.$$

3.1.5 Identification of the Material Parameters

The determination of the coefficients S, ε_{PD}, and D_{1c} that characterize the damage, together with σ_f and σ_u related to plasticity, must be worked out for each material and temperature, from experiments performed on those cases in which the damage is easiest to measure. This involves:

— damage as uniform as possible;
— tensile experiment;
— measurement by means of elasticity change as explained in Section 1.3.2.

This restricts us to monotonic tensile experiment strain imposed at a constant strain rate, or very low cycle fatigue at a constant amplitude of strain.
Let us assume that a good tensile test has been performed with measurement of the damage during unloading by elasticity change (Figure 3.4):
For the one-dimensional case:

$$Y = \frac{\tilde{\sigma}_{eq}^2 R_\nu}{2E} = \frac{\sigma^2}{2E(1-D)^2} \text{ as } R_\nu = 1,$$

$$\dot{p} = \left(\frac{2}{3}\dot{\varepsilon}_{ij}^p \dot{\varepsilon}_{ij}^p\right)^{1/2} = |\dot{\varepsilon}_p| \text{ and in monotonic loading } p = \varepsilon_p,$$

$$\dot{D} = \frac{\sigma^2}{2ES(1-D)^2} |\dot{\varepsilon}_p| \cdot H_{(\varepsilon_p - \varepsilon_{PD})}.$$

— ε_{PD} is the plastic strain below which there is no appreciable damage (see Figure 3.4).
— σ_u is the classical ultimate stress as shown in Figure 3.4. It is also interesting to know the yield stress σ_y as it may help to make a better choice of the plastic threshold σ_s when needed.
— σ_f is the fatigue limit taken, as already mentioned in Section 3.1.2, as the stress amplitude corresponding to a number of cycles to failure in a tension

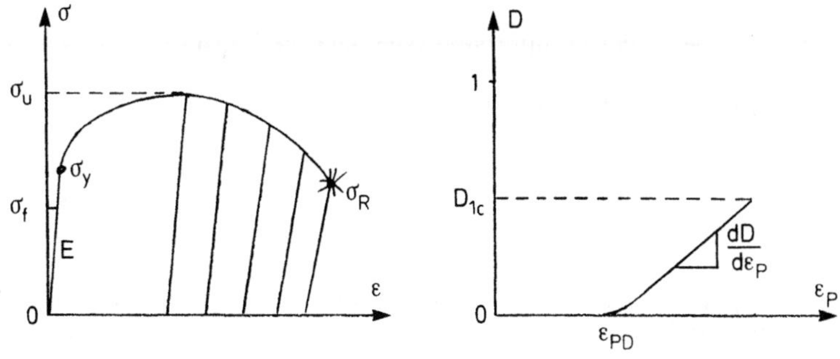

Fig. 3.4. Identification of damage coefficients

compression fatigue test of about 10^6 to 10^7. Usually:

$$\frac{2}{3}\sigma_y < \sigma_f < \sigma_y.$$

— The last parameter S is determined from the slope of the curve: damage D versus the plastic strain ε_p:

$$\dot{D} = \frac{\sigma^2}{2ES(1-D)^2}\dot{\varepsilon}_p$$

or

$$\frac{dD}{d\varepsilon_p} = \frac{\sigma^2}{2ES(1-D)^2}.$$

At each point of the curve, D is known, σ is known from the stress strain curve, $dD/d\varepsilon_p$ is estimated and E is known from a previous identification:

$$S = \frac{\sigma^2}{2E(1-D)^2\dfrac{dD}{d\varepsilon_p}}.$$

Several points may be considered in order to obtain S as the best average.
The main difficulty involved in this identification lies in obtaining a good stress strain curve in the softening range where necking occurs. To avoid any instability, a "soft" machine is required with an excellent feedback system, which can accurately impose the strain in the damaged region of the specimen. This strain must be measured locally by a small strain gauge (see Section 1.3).

3.1.6 Exercise on Identification by a Low Cycle Test

A way to avoid, at least partly, the problem of localization by necking is to perform a tension compression test at constant amplitude of strain imposed at such a value that the strain-hardening is saturated at the first cycle or after a few cycles.
The general kinetic damage law in one dimension is written as:

$$\dot{D} = \frac{\tilde{\sigma}^2}{2ES}|\dot{\varepsilon}_p|\, H(\varepsilon_p - \varepsilon_{pD}),$$

with
$$\tilde{\sigma} = \frac{\sigma}{1-D} \text{ under tension}$$

and
$$\tilde{\sigma} = \frac{\sigma}{1-Dh} \text{ under compression (see Section 2.3.4),}$$

with the condition for crack initiation being,

$$D_c = D_{1c}\frac{\sigma_u^2}{\tilde{\sigma}^{*2}} = D_{1c},$$

as
$$R_\nu = 1 \text{ and } \frac{\sigma_{eq}}{1-D} = \sigma_u.$$

Determination of S, p_D, D_c, and h from the experimental result shown in Figure 3.5 for a particular material.
1) *Plot the evolution of the elasticity modulus (divided by Young's modulus) in tension and compression as a function of the number of cycles* (Figure 3.6).

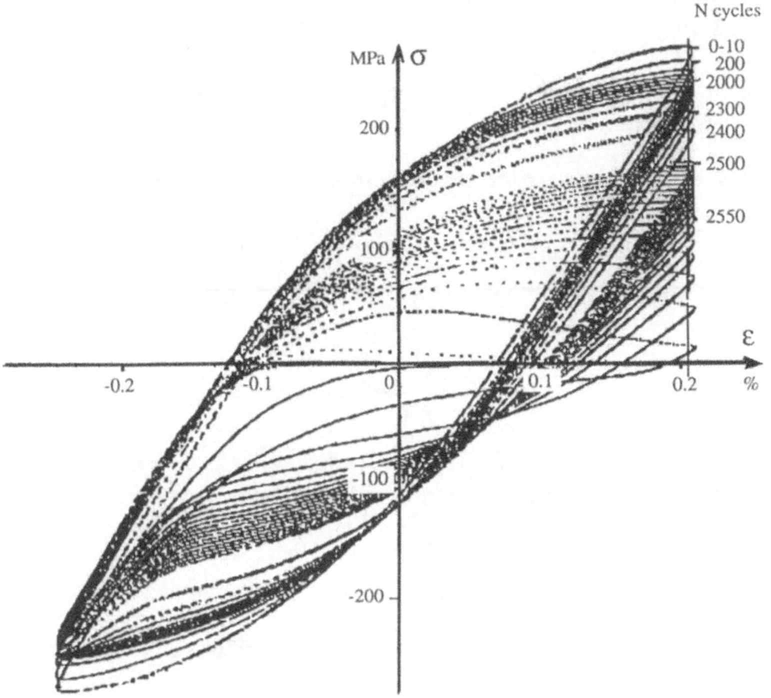

Fig. 3.5. Low cycle fatigue stress strain pattern, for AISI 316 L stainless steel (after J. Dufailly)

From these curves, deduce the crack closure coefficient h,

$$h = \frac{1 - \tilde{E}^-/E}{1 - \tilde{E}^+/E} = 0.2,$$

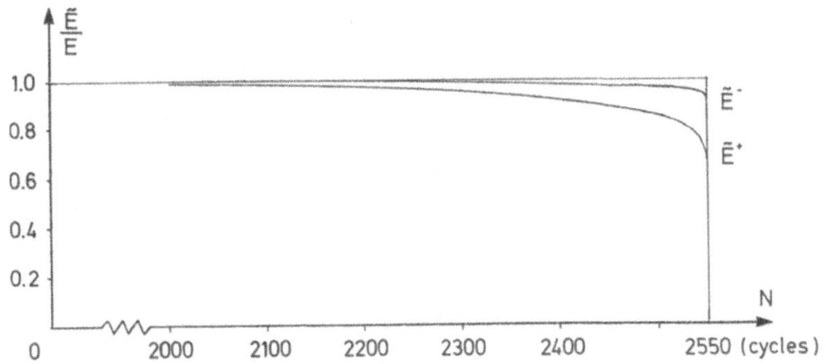

Fig. 3.6. Elasticity modulus evolutions

taken as the mean value for several values of N. *Also deduce the damage as a function of N by*

$$D = 1 - \frac{\tilde{E}^+}{E}, \quad \text{or} \quad D = \frac{1}{h}\left[1 - \frac{\tilde{E}^-}{E}\right].$$

2) *Calculate the accumulated plastic strain as a function of the number of cycles.* For one cycle i,

$$\Delta\varepsilon_{pi} = \Delta\varepsilon - \Delta\varepsilon_{ei} = \Delta\varepsilon - \left(\frac{\sigma_{Mi}}{\tilde{E}_i^+} - \frac{\sigma_{mi}}{\tilde{E}_i^-}\right),$$

$$p = 2\sum_{i=1}^{N} \Delta\varepsilon_{pi}.$$

Plot the damage as a function of p (Figure 3.7.).
This graph allows for the determination of p_D and D_c:

$$p_D = \text{Max}(p_{(D=0)}) \simeq 6,$$
$$D_c = \text{Max}(D_{(p)}) \simeq 0.15.$$

3) *Determine the parameter S* as the average value of several computations such as:

$$S = \frac{\sigma_M^2}{2E(1-D)^2 \dfrac{dD}{dp}}, \quad S \simeq 7Mp_a.$$

3.2 Brittle Damage of Metals, Ceramics, Composites and Concrete

Damage is generally termed brittle when it occurs by decohesion without any sensible plastic strain at the mesoscale. Damage mechanics, as developed, always

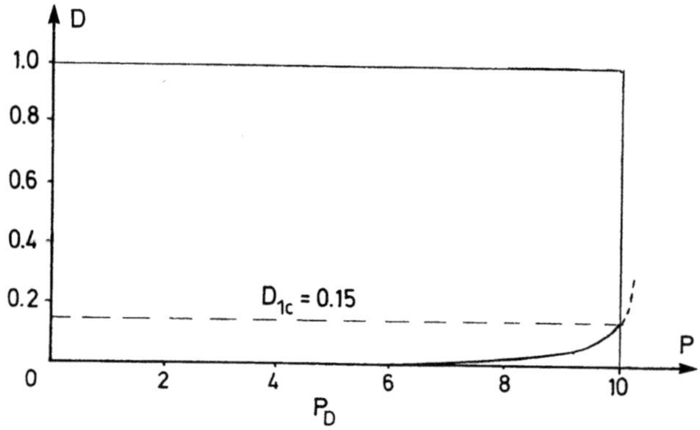

Fig. 3.7. Damage evolution for a low cycle fatigue, on AISI 316 stainless steel

associates damage growth with plastic dissipation: $\dot{D}$ is proportional to $\dot{p}$. For certain kinds of materials such as ceramics, concrete or high strength quenched steels, there is no measurable plastic strain at the mesoscale up to failure. Nevertheless, although p may be considered to be zero at this scale, at the microscale, where the damage develops as a localized phenomenon, some plastic strain always occur in these very small damaged volumes: slips at grain boundaries or interfaces, movements of dislocations in neighborhood of defects in metals.
In cases in which these permanent micro strains may be neglected, the damage is refered to as ''pure brittle''; otherwise, it is called ''quasi brittle''.

3.2.1 Pure Brittle Damage

As there is no plastic strain ($\dot{p} = 0$) when brittle damage is involved, from the general law of kinetic damage, the damage is also zero. Coming back to the instability criterion of Section 3.1.3. the pure brittle rupture occurs at the same time as the instability, that is, for:

$$Y = Y_c = \frac{\sigma_\infty^2}{2E} \simeq \frac{\sigma_u^2}{2E}$$

or, with $D = 0$,

$$Y = \frac{\sigma^{*2}}{2E} = \frac{\sigma_u^2}{2E}$$

or

$$\boxed{\sigma^* = \sigma_{eq} R_\nu^{1/2} = \sigma_u}$$

This equation shows the strong influence of the triaxiality ratio contained in R_ν:

$$R_\nu = \frac{2}{3}(1+\nu) + 3(1-2\nu)\left(\frac{\sigma_H}{\sigma_{eq}}\right)^2.$$

The characteristics of the material are expressed by σ_u; this is the stress to rupture in a one-dimensional tension test.

3.2.2 Quasi-Brittle Damage

When the behavior is brittle at the mesoscale but localized damage growth occurs at the microscale, we are dealing with quasi-brittle damage.
Consider once again a mesovolume element exhibiting elastic behavior everywhere except in a small microvolume μ representing a weak defect subjected to elasticity, plasticity and damage (Figure 3.8).
The matrix is elastic with a yield stress σ_y and a fatigue limit σ_f. The inclusion has the same properties as the matrix except that it is perfectly plastic with a plastic threshold σ_s^μ and a fatigue limit σ_f^μ. Its weakness comes from the value of the plastic threshold, which may be taken equal to the fatigue limit of the material, as it is the lowest stress giving rise to possible damage:

$$\sigma_s^\mu = \sigma_f.$$

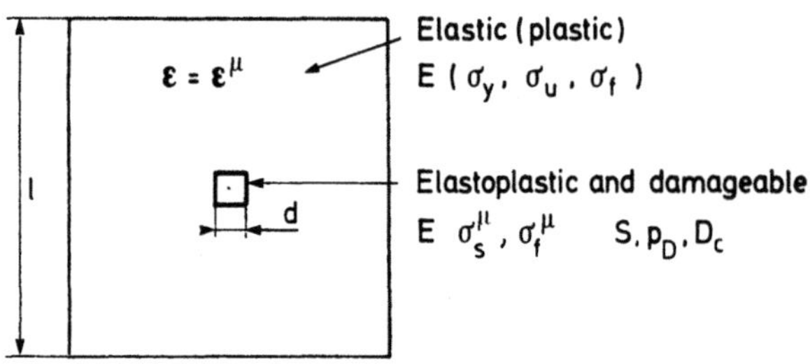

Fig. 3.8. Two-scale volume element

Furthermore, the weakness also comes from the fatigue limit σ_f^μ assumed to be reduced in the same proportion as the plastic threshold:

$$\sigma_f^\mu = \sigma_f \frac{\sigma_s^\mu}{\sigma_y}.$$

The complete inclusion problem will be solved numerically in Section 4.3. by "locally coupled analysis". Here, however, some approximations allow us to derive the rupture conditions without resolution of the complete set of constitutive equations. Starting with the kinetic damage law for the inclusion,

$$\dot{D} = \frac{Y^\mu}{S}\dot{p}^\mu,$$

we wish to express Y^μ and $\dot{p}^\mu$ as functions of macroscopic quantities such as the elastic strain $\boldsymbol{\varepsilon}$ and the stress $\boldsymbol{\sigma}$.

According to the Lin-Taylor hypothesis, we may assume that the inclusion is subjected to the state of strain (or strain rate) of the matrix, which is taken to be uniform:

$$\dot{\boldsymbol{\varepsilon}}^\mu = \dot{\boldsymbol{\varepsilon}}.$$

Neglecting the elastic strain $\boldsymbol{\varepsilon}^{e\mu}$ in comparison to the plastic strain $\boldsymbol{\varepsilon}^{p\mu}$ in the inclusion allows us to write:

$$\dot{p}^\mu = \left(\frac{2}{3}\dot{\varepsilon}_{ij}^{p\mu}\dot{\varepsilon}_{ij}^{p\mu}\right)^{1/2} = \left(\frac{2}{3}\dot{\varepsilon}_{ij}^{\mu D}\dot{\varepsilon}_{ij}^{\mu D}\right)^{1/2} = \left(\frac{2}{3}\dot{\varepsilon}_{ij}^{D}\dot{\varepsilon}_{ij}^{D}\right)^{1/2} = \dot{\varepsilon}_{eq}.$$

if the yield criterion is satisfied.

Next, what about Y^μ?

$$Y^\mu = \frac{\sigma_{eq}^{\mu^2} R_\nu^\mu}{2E(1-D)^2}.$$

The inclusion being perfectly plastic, then, from the yield criterion,

$$\frac{\sigma_{eq}^\mu}{1-D} = \sigma_s^\mu.$$

The triaxiality factor may also be expressed as a function of the stress at the mesoscale:

$$R_\nu^\mu = \frac{2}{3}(1+\nu) + 3(1-2\nu)\left(\frac{\sigma_H^\mu}{\sigma_{eq}^\mu}\right)^2,$$

$$\sigma_{eq}^\mu = (1-D)\sigma_s^\mu;$$

$\sigma_H^\mu = \dfrac{E(1-D)}{1-2\nu}\varepsilon_H^{e\mu}$ from the damaged elasticity at the microscale,

but $\varepsilon_H^{e\mu} = \varepsilon_H$ at the mesoscale since $\boldsymbol{\varepsilon}^{e\mu} + \boldsymbol{\varepsilon}^{p\mu} = \boldsymbol{\varepsilon}$ and $t_r(\boldsymbol{\varepsilon}^{p\mu}) = 0$;

$\varepsilon_H = \dfrac{1-2\nu}{E}\sigma_H$ from pure elasticity at the mesoscale;

then $\sigma_H^\mu = (1-D)\sigma_H$ and $\dfrac{\sigma_H^\mu}{\sigma_{eq}^\mu} = \dfrac{\sigma_H}{\sigma_s^\mu}$.

The damage threshold $p_D = \varepsilon_{pD}\dfrac{\sigma_u - \sigma_f}{\sigma_{eq} - \sigma_f}$ becomes $p_D = \varepsilon_{pD}\dfrac{\sigma_u - \sigma_f}{\sigma_s^\mu - \sigma_f^\mu}$ because for the inclusion, $\sigma_{eq} = \sigma_s^\mu$ and $\sigma_f = \sigma_f^\mu$, the fatigue limit remaining σ_f for the case of reference.

It is not possible to write the yield criterion exactly in terms of macroscopic quantities. The following approximation is made:

$$\dot{D} = 0 \quad \text{if} \quad \sigma_{eq} < \sigma_f .$$

Finally,

$$\boxed{\dot{D} = \frac{\sigma_s^{\mu^2}}{2ES}\left[\frac{2}{3}(1+\nu) + 3(1-2\nu)\left(\frac{\sigma_H}{\sigma_s^\mu}\right)^2\right]\dot{\varepsilon}_{eq}} \quad \text{if } \varepsilon_{eq} \geq p_D, \text{ and if } \sigma_{eq} > \sigma_f$$

with

$$\cdot\, p_D = \varepsilon_{pD}\frac{\sigma_u - \sigma_f}{\sigma_s^\mu - \sigma_f^\mu}, \quad \sigma_f^\mu = \sigma_f\frac{\sigma_s^\mu}{\sigma_y}, \quad \sigma_s^\mu = \sigma_f .$$

caution is advised here: $\dot{\varepsilon}_{eq} = \left(\frac{2}{3}\dot{\varepsilon}_{ij}\dot{\varepsilon}_{ij}\right)^{1/2}$ and not $\frac{d}{dt}\left(\frac{2}{3}\varepsilon_{ij}\varepsilon_{ij}\right)^{1/2}$!

The integration of this damage rate equation may be performed for the following initial and final conditions:

$$\varepsilon_{eq} < p_D \rightarrow D = 0,$$

$$\varepsilon_{eq} = \varepsilon_{eq}^R \leftarrow D = D_c = D_{1c}\frac{\sigma_u^2}{\sigma_s^{\mu^2} R_\nu}.$$

This last condition is, strictly speaking, a crack initiation at the microscale but can be shown to also correspond to the brittle fracture of the matrix ! The strain energy release rate at the mesoscale corresponding to a microcrack of surface d^2 is

$$G = -\frac{1}{2}\frac{\delta W}{\delta A}\bigg|_{\sigma=\text{const}} .$$

But $-\frac{1}{2}\delta W|_{\sigma=\text{const}}$ is also equal to the energy dissipated in the inclusion by the damaging process:

$$G_{(\delta A)} = \frac{d^3 \int_0^{D_c} Y\, dD}{\delta A}.$$

Assuming a constant strain energy density release rate $Y = Y_c$ and a critical value of the damage $D_c = 1$ yields

$$G_{(\delta A)} = Y_c d.$$

Using the result of Section 2.3.1 in which a simple relation was found between the critical value of the strain energy density release rate Y_c and the dimension of the crack (l in Chapter 2, d here), and the toughness G_c,

$$Y_c d = G_c$$

shows that $\boxed{G(D = D_c \text{ at the microscale}) = G_C \text{ (at the mesoscale).}}$

We can conclude that condition of crack initiation at the microscale satisfies also the brittle crack initiation criterion at the mesoscale.

In the case of constant hydrostatic stress loading, the integration of the damage rate equation becomes obvious:

Calling $$R_{\nu\left(\frac{\sigma_H}{\sigma_s^\mu}\right)} = \frac{2}{3}(1+\nu)+3(1-2\nu)\left(\frac{\sigma_H}{\sigma_s^\mu}\right)^2 = \text{const.},$$

$$D_c = \frac{\sigma_f^2}{2ES} R_{\nu\left(\frac{\sigma_H}{\sigma_s^\mu}\right)} \left(\varepsilon_{eq}^R - p_D\right)$$

or, with the assumption, $\sigma_s^\mu = \sigma_f$,

$$\varepsilon_{eq}^R = \varepsilon_{PD} \frac{\sigma_u - \sigma_f}{\sigma_f - \dfrac{\sigma_f^2}{\sigma_y}} + \frac{2ESD_c}{\sigma_f^2 R_{\nu\left(\frac{\sigma_H}{\sigma_s^\mu}\right)}}.$$

This result accounts for any kind of constant hydrostatic stress loading: monotonic, fatigue, etc., with the proper values of ε_{p_D} and D_c; remember, however, that ε_{eq} is the accumulated total strain.

3.2.3 Exercise on the Influence of the Triaxiality on Rupture

In practical problems of engineering the triaxiality ratio varies from 0 (pure shear) to 5 or 6 at the tips of very sharp notches, $\dfrac{\sigma_H}{\sigma_{eq}} = \dfrac{1}{3}$ being the value for pure tension.

For pure brittle damage, plot the von Mises equivalent stress at failure divided by the one-dimensional damage equivalent stress as a function of the triaxiality ratio. Take Poisson's ratio as a parameter: $\nu = 0.2$, $\nu = 0.3$, $\nu = 0.4$ (Figure 3.9).

— three-dimensional failure: $\sigma^* = \sigma_{eq} R_\nu^{1/2} = \sigma_u$;
— one-dimensional failure: $\sigma^* = \sigma_R = \sigma_u$.

This exercise should convince you that using the von Mises criterion $\sigma_{eq} = \text{const.}$ for failure is like using a hammer to open a bottle of wine!

For quasi-brittle damage, plot versus the triaxiality at microscale $\dfrac{\sigma_H^\mu}{\sigma_{eq}^\mu}$ the equivalent strain to rupture (minus p_D) divided by the one-dimensional strain to rupture (minus ε_{PD}) for the three values of Poisson's ratio $\nu = 0.2, 0.3, 0.4$. For simplicity take $D_c = D_{1c}$, and $\sigma = 3\sigma_s^\mu$ for the one-dimensional case at the mesoscale which induces a three-dimensional state of stress at the microscale (Figure 3.10).

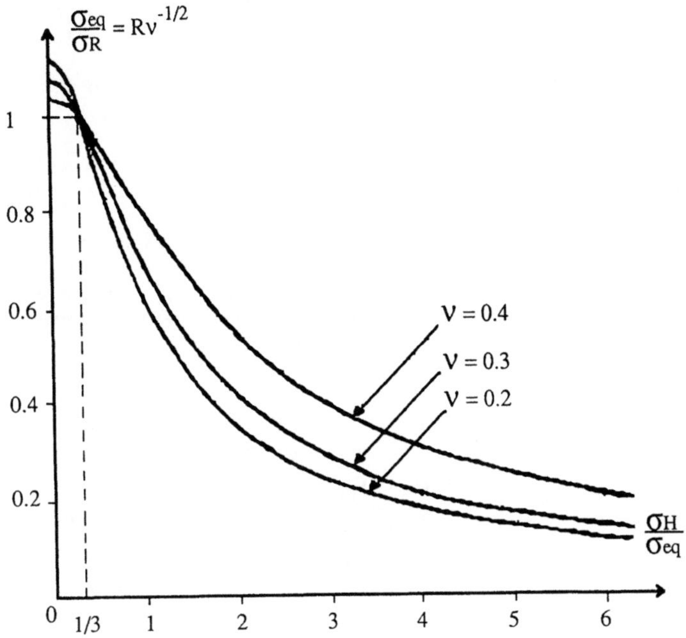

Fig. 3.9. Stress criterion of brittle failure

3.3 Ductile and Creep Damage of Metals and Polymers

Recalling Section 1.1.4, the damage process is called ductile or creep when it occurs simultaneously with large permanent strains. At the microscale it mainly involves the nucleation and growth of cavities causing ductile damage in metals and polymers; in metals submitted to elevated temperature, it causes creep damage on the form of grain boundary sliding and decohesion.

3.3.1 Ductile Damage

The general damage law is straightforward in its application to ductile damage:

$$\dot{D} = \frac{Y}{S}\dot{p}H(p - p_D).$$

The damage evolution may be calculated for any process of loading defined by the histories of the accumulated strain rate $\dot{p}(t)$, the von Mises equivalent stress $(\sigma_{eq}(t))$ and the hydrostatic stress $\sigma_H(t)$.

Case of proportional loading

This is a particular case of loading which often occurs in structures for which integration is simple.

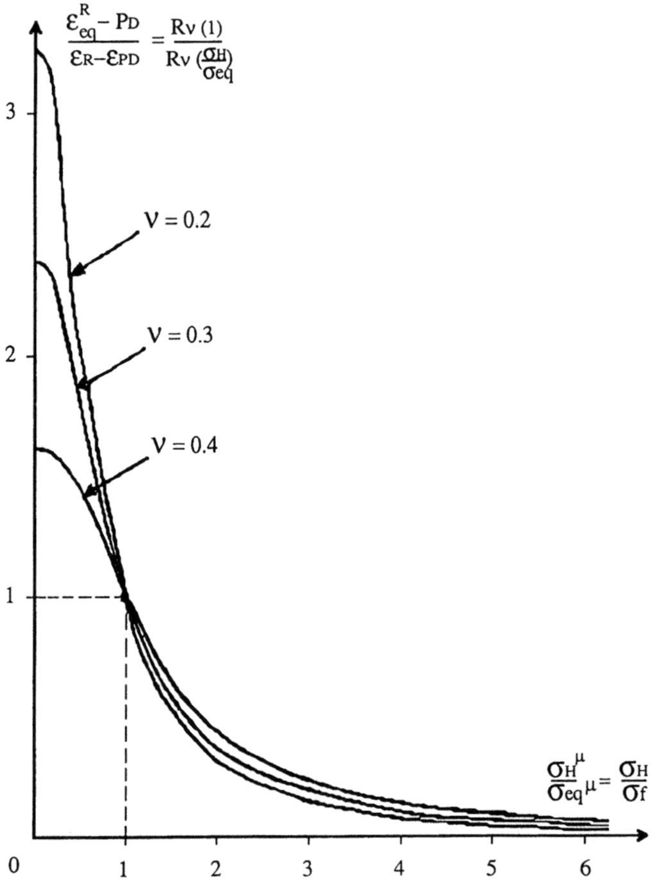

Fig. 3.10. Strain criterion for quasi-brittle failure

The following results were obtained in Section 2.2.4. Starting with the definition of a proportional loading in a structure:

$$\sigma_{(M,t)} = \mathbf{S}_{(M)} \cdot \tau_{(t)};$$

it follows that

$$\frac{\sigma_H}{\sigma_{eq}} = \text{const.},$$

$$R_\nu = \text{const.}$$

Assuming as in Section 3.1.4, that the ductile damage occurs when the strain-hardening is saturated or almost saturated, then from the plasticity criterion:

$$\tilde{\sigma}_{eq} - \sigma_s = 0 \quad \text{or} \quad \frac{\sigma_{eq}}{1-D} = \sigma_s.$$

This approximation makes the strain energy density release rate a constant:

$$Y = \frac{\sigma_{eq}^2 R_\nu}{2E(1-D)^2} = \frac{\sigma_s^2 R_\nu}{2E} = \text{const}.$$

and the integration of the damage constitutive equation becomes obvious:

$$D = \int_0^t \frac{\sigma_s^2 R_\nu}{2ES} \dot{p}(t) H(p - p_D) = \frac{\sigma_s^2 R_\nu}{2ES} \int_{p_D}^{p} dp;$$

$$\boxed{D = \frac{\sigma_s^2 R_\nu}{2ES} (p - p_D).}$$

This makes it possible to determine the value of the damage at each point of a structure as a function of the accumulated plastic strain, provided that the loading is proportional.

In one dimension and monotonic loading:

$$R_\nu = 1,$$

$$p = \varepsilon_p,$$

$$D_c = D_{1c} \frac{\sigma_u^2}{\sigma_s^2};$$

the three dimensional equation becomes

$$D = \frac{\sigma_s^2}{2ES} (\varepsilon_p - \varepsilon_{p_D}).$$

Writing this for the critical condition of crack initiation,

$$D = D_c \rightarrow \varepsilon_p = \varepsilon_{p_R},$$

$$D_c = \frac{\sigma_s^2}{2ES} (\varepsilon_{p_R} - \varepsilon_{p_D}).$$

This allows us to write the one-dimensional evolution of damage as

$$D = D_c \frac{\varepsilon_p - \varepsilon_{p_D}}{\varepsilon_{p_R} - \varepsilon_{p_D}},$$

which is the equation of a straight line lying between the two points ($D = 0$, $\varepsilon_p = \varepsilon_{p_D}$) and ($D = D_c, \varepsilon_p = \varepsilon_{p_R}$) as shown in Figure 3.11.

3.3.2 Exercises on the Fracture Limits in Metal Forming

Metal forming is a process by which a component of a given shape is obtained from a cast product by means of large plastic deformations: rolling, forging, deep drawing, extrusion, etc. The state of strain to be obtained is limited by two phenomena: the plastic instability giving rise to unexpected local changes of shape, and the damage, which may produce unexpected cracks. Let us analyze the later phenomenon.

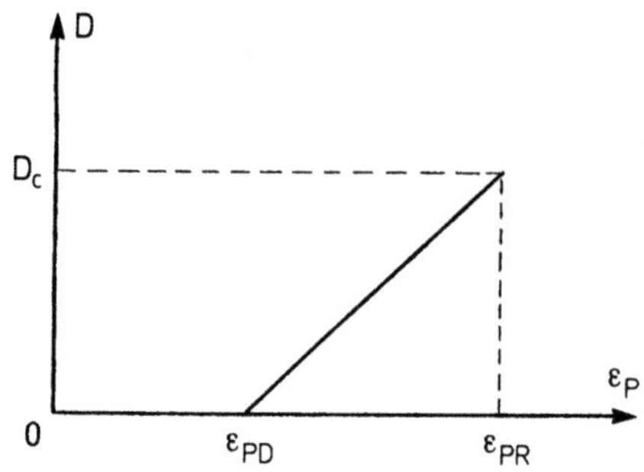

Fig. 3.11. Monotonic tensile damage evolution

Master curves of ductile fracture

Calculation of the maximum accumulated plastic strain p_R corresponding to the critical value of damage as a function of the triaxiality ratio

— using the general damage law:

$$\dot{D} = \frac{Y}{S}\dot{p}H(p-p_D).$$

— in the case of proportional loading:

$$\frac{\sigma_H}{\sigma_{eq}} = \text{constant during the entire the process,}$$

— for the critical value of the damage: $D_c = D_{1c} = 1$.

Integrating the kinetic law once in the three-dimensional case and once in the one-dimensional case gives the result:

$$\frac{p_R - p_D}{\varepsilon_{p_R} - \varepsilon_{p_D}} = \frac{1}{R_\nu} = \left[\frac{2}{3}(1+\nu)+3(1-2\nu)\left(\frac{\sigma_H}{\sigma_{eq}}\right)^2\right]^{-1},$$

which may be plotted on the graph $\left(\frac{p_R - p_D}{\varepsilon_{p_R} - \varepsilon_{p_D}}\right), \left(\frac{\sigma_H}{\sigma_{eq}}\right)$ as shown in Figure 3.12, which is similar to Figure 3.10. The one-dimensional case is of course represented by the point

$$\left(\frac{\sigma_H}{\sigma_{eq}} = \frac{1}{3}, \quad \frac{p_R - p_D}{\varepsilon_{p_R} - \varepsilon_{p_D}} = 1\right).$$

These limit curves depend upon the type of material by means of p_D and slightly by means of Poisson's ratio if $0.25 \leq \nu \leq 0.33$. They strongly depend upon the triaxiality ratio. The ductility measured by the maximum accumulated plastic strain at crack initiation decreases substantially with the triaxiality ratio; it is well known that a "sharp notch makes the material more brittle". (In fact it is not the material itself but the effect of the hydrostatic loading on it).

Fracture limits in the deep drawing process (J.P. Cordebois 1983)

Deep drawing consists of indenting a sheet of metal to obtain a three-dimensional form (Figure 3.13). In practice it is interesting to know a priori the limits in the strain space between the states of possible strains and the states of strain that do not occur without a crack.

Calculation of the relation between the principal strain components corresponding to the critical value of damage may be performed:

— in the case of plane stresses;
— neglecting elastic strain;
— using the general damage law with $\varepsilon_{PD} = 0$ and $D_c = 1$;
— assuming proportional loading;
— neglecting the kinematic hardening in the plastic constitutive equation.

From the master curve of ductile fracture with $p_D = 0$, as $\varepsilon_{PD} = 0$,

$$p_R = \varepsilon_R \left[\frac{2}{3}(1+\nu) + 3(1-2\nu)\left(\frac{\sigma_H}{\sigma_{eq}}\right)^2 \right]^{-1}.$$

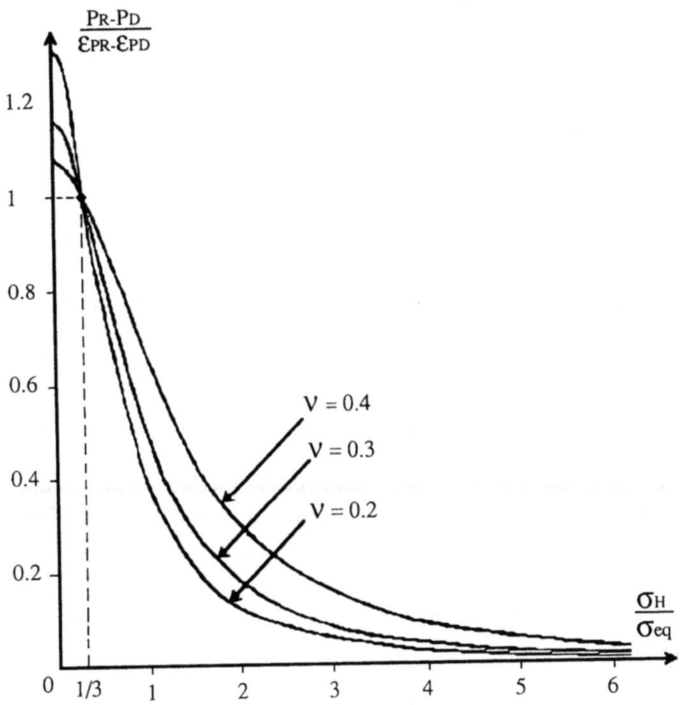

Fig. 3.12. Master curves of ductile fracture limits

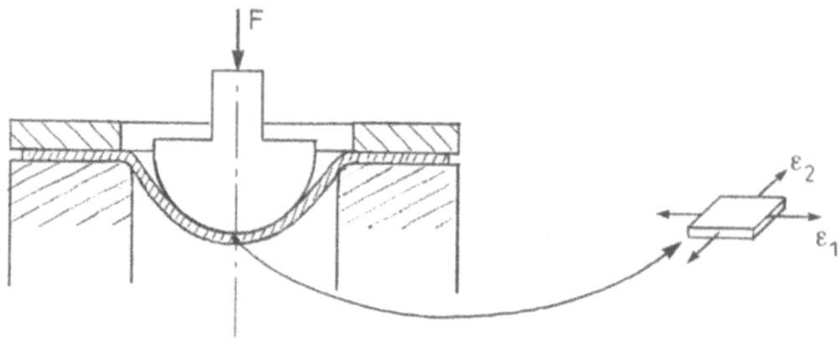

Fig. 3.13. Elementary deep-drawing process

p_R may be calculated as a function of the strain by its definition together with $\varepsilon^p = \varepsilon$:

$$p = \left(\frac{2}{3}\varepsilon_{ij}\varepsilon_{ij}\right)^{1/2}.$$

The plane stress condition imposes for the principal stress components that:

$$[\sigma] = \begin{bmatrix} \sigma_1 & 0 & 0 \\ 0 & \sigma_2 & 0 \\ 0 & 0 & 0 \end{bmatrix},$$

from which the plastic incompressible condition is written for $t_r([\varepsilon]) = 0$:

$$[\varepsilon] = \begin{bmatrix} \varepsilon_1 & 0 & 0 \\ 0 & \varepsilon_2 & 0 \\ 0 & 0 & -(\varepsilon_1 + \varepsilon_2) \end{bmatrix};$$

then

$$p_R = \frac{2}{\sqrt{3}}\left(\varepsilon_1^2 + \varepsilon_2^2 + \varepsilon_1\varepsilon_2\right)^{1/2}.$$

The triaxiality ratio σ_H/σ_{eq} may also be calculated as a function of the strain from the plastic constitutive equations.
From Section 2.1.4,

$$\dot{\varepsilon}_{ij}^p = \frac{3}{2}\frac{\left(\tilde{\sigma}_{ij}^D - X_{ij}^D\right)}{\left(\tilde{\boldsymbol{\sigma}}^D - \mathbf{X}^D\right)_{eq}}\frac{\dot{\lambda}}{1-D}.$$

Assuming no kinematic hardening ($\mathbf{X} = 0$) and proportional loading, it is straightforward to demonstrate the proportionality between strains and components of the deviatoric stress:

$$\frac{\varepsilon_2}{\varepsilon_1} = \frac{\sigma_2^D}{\sigma_1^D},$$

from which it follows that

$$\frac{\sigma_H}{\sigma_{eq}} = \frac{\dfrac{\varepsilon_2}{\varepsilon_1} + 1}{\sqrt{3}\left[\left(\dfrac{\varepsilon_2}{\varepsilon_1}\right)^2 + \dfrac{\varepsilon_2}{\varepsilon_1} + 1\right]^{1/2}}.$$

Then replacing p_R and σ_H/σ_{eq} in the first equation gives the result:

$$\left(\frac{\varepsilon_1}{\varepsilon_R}\right)^2 + \frac{\varepsilon_1\varepsilon_2}{\varepsilon_R^2} + \left(\frac{\varepsilon_2}{\varepsilon_R}\right)^2 - \frac{3}{4}\left[\frac{2}{3}(1+\nu) + (1-2\nu)\frac{\left(\frac{\varepsilon_2}{\varepsilon_1}+1\right)^2}{\left(\frac{\varepsilon_2}{\varepsilon_1}\right)^2 + \frac{\varepsilon_2}{\varepsilon_1} + 1}\right]^{-2} = 0.$$

A plot of $\varepsilon_2/\varepsilon_R$ as a function of $\varepsilon_1/\varepsilon_R$ is the classical failure limit in deep drawing of sheets (Figure 3.14). As for the master curve of ductile fracture, the curves depend slightly upon Poisson's ratio but the result for each particular case depends strongly upon the type of material by means of its characteristic ε_R.

3.3.3 Creep Damage

The application of the kinetic damage law to creep damage is also straightforward. It differs from the case of ductile damage only in that the accumulated plastic strain rate comes from a visco plastic constitutive equation:

$$\dot{D} = \frac{Y}{S}\dot{p}H(p - p_D),$$

in which Y may possibly take into account the microcrack closure effect (see Section 3.1.4) Remember that the material parameter S depends upon temperature. From Section 2.1.4,

$$\dot{p} = \text{Ln}\left[1 - \frac{(\tilde{\boldsymbol{\sigma}}^D - \mathbf{X}^D)_{eq} - R - \sigma_y}{K_\infty}\right]^{-n},$$

with
$$\dot{p} = \frac{\dot{\lambda}}{1-D}.$$

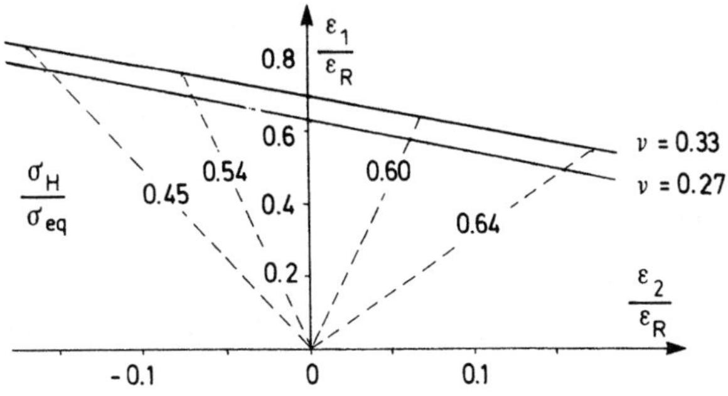

Fig. 3.14. Limit curves of deep drawing

For simplicity, let us consider a simple perfect visco plastic Norton's law with neither strain hardening nor yield stress (see Section 2.1.3). Then the visco plastic multiplier reduces to:

$$\frac{\dot{\lambda}}{1-D} = \dot{p} = \left[\frac{\sigma_{eq}}{K_v(1-D)}\right]^N,$$

where K_v and N are material parameters which can be found for some materials from the table in Section 3.5. Then

$$\dot{D} = \frac{\sigma_{eq}^2 R_\nu}{2ES(1-D)^2}\left[\frac{\sigma_{eq}}{K_v(1-D)}\right]^N H(p-p_D)$$

or

$$\dot{D} = \frac{\sigma_{eq}^{N+2} R\nu}{2ESK_v^N(1-D)^{N+2}} H(p-p_D).$$

The damage becomes a time dependent phenomenon like the visco plastic strain. In the one-dimensional case this constitutive equation reduces to the Kachanov's earlier model:

$$\boxed{\dot{D} = \left[\frac{\sigma}{A(1-D)}\right]^{N+2} H(p-p_D),}$$

with

$$A = (2ESK_v^N)^{\frac{1}{N+2}}.$$

Let us calculate the evolution of the damage as a function of the time in a simple creep process where σ = const.
Let t^* be the time needed to reach $\varepsilon_p = p_D$ by creep without any damage,

$$dD = \left(\frac{\sigma}{A}\right)^{N+2} (1-D)^{-(N+2)}\, dt$$

or

$$\int_0^D (1-D)^{N+2}\, dD = \left(\frac{\sigma}{A}\right)^{N+2} \int_{t^*}^t dt,$$

which leads to

$$D = 1 - \left[1 - (N+3)\left(\frac{\sigma}{A}\right)^{N+2} (t-t^*)\right]^{\frac{1}{N+3}}.$$

The critical value of the damage defining the rupture of the volume element under tension is obtained by (see Section 3.1.3)

$$D_c = D_{1c}\frac{\sigma_\infty^2}{\tilde{\sigma}^2} \leq 1,$$

but as Norton's creep law is used here, $\sigma_\infty \rightarrow \infty$ at rupture when $\dot{\varepsilon}p \rightarrow \infty$. Then $D_c = 1$ is the proper rupture criterion for this analysis. The time to rupture $t_r(D = 1)$ is

$$t_R - t^* = \frac{1}{N+3}\left(\frac{\sigma}{A}\right)^{-(N+2)}.$$

The time t^* must be calculated from a visco plastic model. For simplicity let us again take the Norton's perfect visco plastic law:

$$\dot{\varepsilon}_P = \left(\frac{\sigma}{K_v}\right)^N$$

or

$$\int_0^{p_D} d\varepsilon_p = \left(\frac{\sigma}{K_v}\right)^N \int_0^{t^*} dt, \quad t^* = p_D \left(\frac{\sigma}{K_\nu}\right)^{-N};$$

$$t_R = p_D \left(\frac{\sigma}{K_v}\right)^{-N} + \frac{1}{N+3}\left(\frac{\sigma}{A}\right)^{-(N+2)} \text{ with } p_D = \varepsilon_{pD}\frac{\sigma_u - \sigma_f}{\sigma - \sigma_f}.$$

Then it is possible to draw the graph of the evolution of the damage as a function of time for constant stress creep processes (Figure 3.15).
Viscoplastic damage behavior must be accurately calculated using the viscoplastic constitutive equations coupled with damage as shown in Section 2.1.4; this is only possible by means of numerical calculations. In order to give a general idea based on analytical results, let us again assume the simple Norton's perfect visco-plastic law and calculate the evolution of strain with time for a simple creep process at $\sigma = $ const.:

$$\varepsilon_e = \frac{\sigma}{E},$$

$$\dot{\varepsilon}_p = \left[\frac{\sigma}{K_v(1-D)}\right]^N,$$

$$\dot{D} = \left[\frac{\sigma}{A(1-D)}\right]^{N+2} H(\varepsilon_p - p_D);$$

For $\varepsilon_p < p_D$ and with the initial condition $t = 0 \rightarrow \varepsilon_p = 0$:

$$D = 0 \rightarrow \varepsilon_p = \left(\frac{\sigma}{K_v}\right)^N t;$$

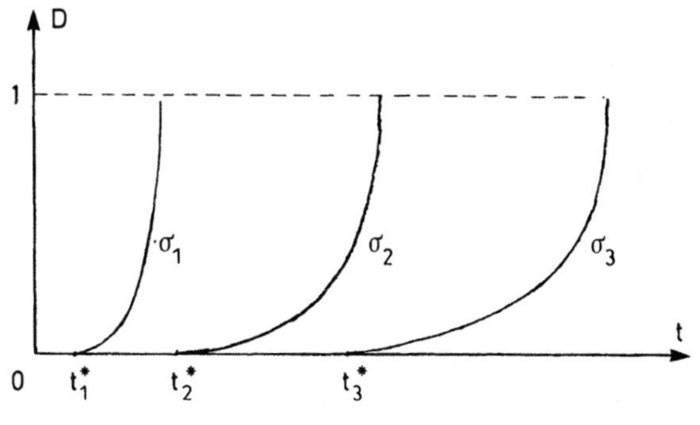

Fig. 3.15. Evolution of the damage in creep processes $\sigma_3 < \sigma_2 < \sigma_1$ (schematic)

For $\varepsilon_p \geq p_D$:

$$\dot{\varepsilon}_p = \left(\frac{\sigma}{K_v}\right)^N (1-D)^{-N}.$$

Taking $(1-D)$ from the preceding calculation gives for $t > t*$

$$\dot{\varepsilon} = \left(\frac{\sigma}{K_v}\right)^N \left[1-(N+3)\left(\frac{\sigma}{A}\right)^{N+2}(t-t^*)\right]^{-\frac{N}{N+3}},$$

$$\varepsilon_p - p_D = \left(\frac{\sigma}{K_v}\right)^N \int_{t^*}^{t} \left[1-(N+3)\left(\frac{\sigma}{A}\right)^{N+2}(t-t^*)\right]^{-\frac{N}{N+3}} dt$$

or

$$\varepsilon_p = p_D + \left(\frac{\sigma}{K_v}\right)^N \frac{1-\left[1-(N+3)\left(\frac{\sigma}{A}\right)^{N+2}(t-t^*)\right]^{\frac{3}{N+3}}}{3\left(\frac{\sigma}{A}\right)^{N+2}}.$$

The shape of the graph of the total strain $\varepsilon = \varepsilon_e + \varepsilon_p$ as a function of time for different values of σ is given in Figure 3.16.

3.3.4 Exercise on Isochronous Creep Damage Curves

In engineering practice it is of interest to know the stress which, when applied constantly to a given material at a given temperature over a certain period of time, causes a certain amount of plastic strain or damage. Those isochronous curves may be obtained by interpolation of many experimental results; they can also be calculated from the constitutive equations of plastic strain and damage.

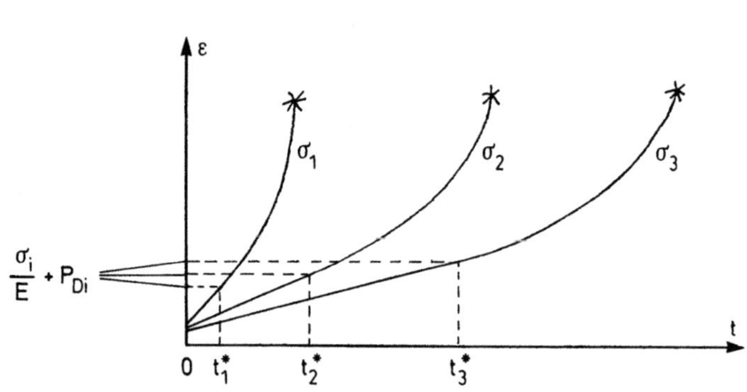

Fig. 3.16. Evolution of the strain up to rupture in creep processes, $\sigma_3 < \sigma_2 < \sigma_1$

Calculation of the time to rupture in pure creep

This calculation may be performed for:

— tension
— in the case of Norton's perfect viscoplasticity for the material AISI 316 at 600°C and constant damage threshold $p_D = \varepsilon_{pD}$:

$$K_\nu = 765 \text{ MPa } h^{1/N} \qquad A = 900 \text{ MPa } h^{\frac{1}{N+2}}$$
$$N = 8.2 \qquad \varepsilon_{p_D} = 2 \cdot 10^{-2}.$$

For $\varepsilon_p < p_D$ the relation between the stress, the plastic strain and the time calculated in Section 3.3.3 is

$$\varepsilon_p = \left(\frac{\sigma}{K_v}\right)^N t \text{ or } t = \varepsilon_p \left(\frac{\sigma}{K_v}\right)^{-N}.$$

Then, it is easy to plot t as a function of σ taking ε_p as a parameter (Figure 3.17). For $\varepsilon_p \geq p_D$, the relation between the stress, the damage and the time is:

$$D = 1 - \left[1 - (N+3)\left(\frac{\sigma}{A}\right)^{N+2} (t - t^*)\right]^{\frac{1}{N+3}},$$

with
$$t^* = p_D \left(\frac{\sigma}{K_v}\right)^{-N}$$

or
$$t = \frac{p_D}{\left(\frac{\sigma}{K_v}\right)^N} + \frac{1 - (1 - D)^{N+3}}{(N+3)\left(\frac{\sigma}{A}\right)^{N+2}}.$$

Taking D as a parameter one may plot t as a function of σ (Figure 3.17). At rupture, the critical value of the damage D_c is

$$D_c = 1,$$

which gives the time to rupture to be plotted on the same graph (Figure 3.17).

3.4 Fatigue Damage

The damage process is called fatigue when the loading is repeated a large number of times with the same amplitude or with varying amplitude. It can be one dimensional, in tension or tension-compression, or three dimensional.

3.4.1 Low Cycle Fatigue

The case of low cycle fatigue occurs when the loading is large enough to produce plastic strain of the order of magnitude of the elastic strain. This means that the number of cycles to failure N_R is relatively low. An order of magnitude is

$$N_R < 10000 \text{ cycles}.$$

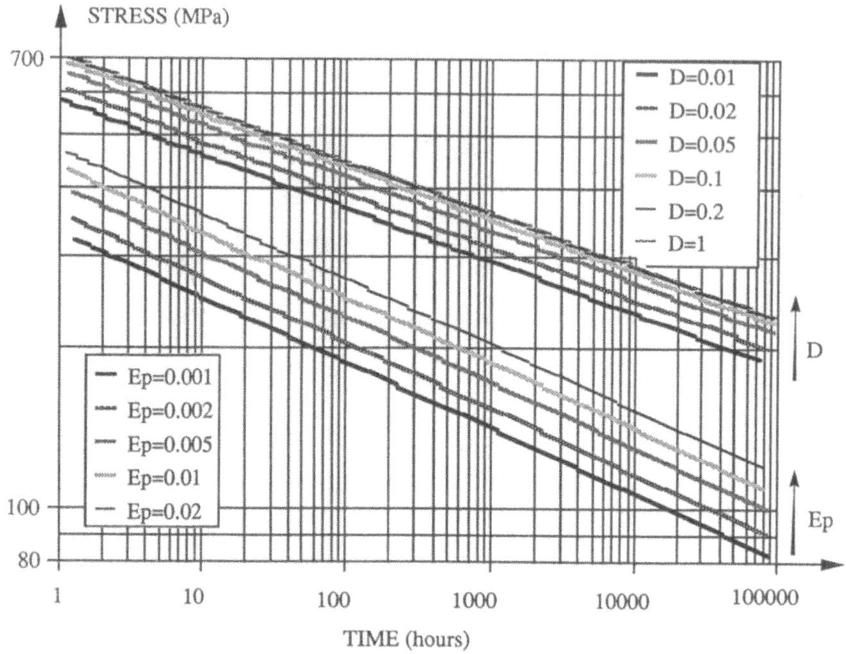

Fig. 3.17. Isochronous creep damage curves (after F. Hild)

One-dimensional periodic fatigue

Let us start with the general kinetic damage law written for the one-dimensional case defined by the stress $\sigma(t)$ and the strain $\varepsilon(t)$ as functions of time t:

$$\dot{D} = \frac{Y}{S}\dot{p} \cdot H(p - p_D),$$

$$Y = \frac{\sigma_{eq}^2 R_\nu}{2E(1-D)^2} = \frac{\sigma^2}{2E(1-D)^2},$$

$$\dot{p} = |\dot{\varepsilon_p}|,$$

$$\dot{D} = \frac{\sigma^2}{2ES(1-D)^2}|\dot{\varepsilon}_p| \text{ if } \varepsilon_p > p_D.$$

The number of cycles to failure is the sum of two parts: the number of cycles N_o to reach p_D for microcrack nucleation, and the number of cycles N_D during which the damage occurs, initiating a crack at the mesoscale:

$$N_R = N_o + N_D.$$

N_o must be calculated by the classical elastoplasticity constitutive equations without any coupling with damage, as during this period $D = 0$.

Let us consider a periodic strain-controlled loading of constant amplitude $\Delta\varepsilon$ (Figure 3.18).
The elastoplastic constitutive equations of Section 2.1.4 show that after some cycles the elasto plastic process is stabilized to an amplitude of stress which remains constant until $p = p_D$:

$$\Delta\varepsilon = \text{const.} \rightarrow \Delta\sigma = \text{const.},$$

as

$$\Delta\varepsilon = \Delta\varepsilon_e + \Delta\varepsilon_p \text{ and } \Delta\varepsilon_e = \Delta\sigma/E$$

$$\Delta\varepsilon_p = \text{const.}$$

The accumulated plastic strain after one cycle is

$$\frac{\delta p}{\delta N} = \int_{1\ \text{cycle}} |\dot{\varepsilon}_p|\, dt = 2\Delta\varepsilon_p,$$

and for N cycles,

$$p_{(N)} = 2N\Delta\varepsilon_p.$$

Then

$$N_o = \frac{p_D}{2\Delta\varepsilon_p}, \text{ with } p_D = \varepsilon_{pD}\frac{\sigma_u - \sigma_f}{\sigma_M - \sigma_f}.$$

The calculation of N_D requires the integration of the kinetic damage law over one cycle and then over N cycles:

$$\dot{D} = \frac{\sigma^2}{2ES(1-D)^2}|\dot{\varepsilon}_p|.$$

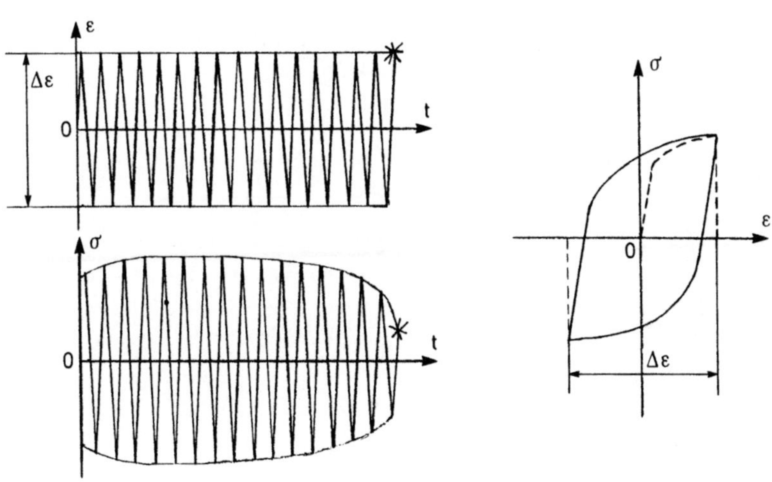

Fig. 3.18. Strain controlled low cycle fatigue

Normally this calculation is performed by computer, as $\frac{\sigma}{1-D}$ is a function of ε_p, but the general trends may be obtained analytically from some simplifications:

- The material is first assumed to be perfectly plastic during one cycle with a threshold $\sigma_s = \frac{\Delta\sigma}{2}$, and the variation of the damage is neglected for the integration over one cycle (Figure 3.19).
- The strain and damage behaviors are assumed to be identical both under tension and compression, this corresponds to the crack closure parameter h being equal to 1 (Section 2.3.4). This allows us to calculate the damage per cycle as:

$$\frac{\delta D}{\delta N} = \int_{1\ \text{cycle}} \dot{D}\,dt = \frac{\left(\frac{\Delta\sigma}{2}\right)^2}{2ES(1-D)^2} 2 \int_0^{\Delta\varepsilon_p} \dot{\varepsilon}_p\,dt = \frac{(\Delta\sigma)^2}{4ES(1-D)^2}\Delta\varepsilon_p .$$

A cyclic relationship between $\Delta\varepsilon_p$ and $\Delta\sigma$ is assumed. It can be derived numerically from the equations of Section 2.1.4 and has strong experimental support:

$$\Delta\varepsilon_p = \left[\frac{\Delta\sigma}{K_p(1-D)}\right]^M .$$

K_p and M are material parameters which can be found for some materials in the table in Section 3.5.

It follows that:

$$\frac{\delta D}{\delta N} = \frac{\left(K_p \Delta\varepsilon_p^{1/M}\right)^2}{4ES}\Delta\varepsilon_p = \frac{K_p^2}{4ES}\Delta\varepsilon_p^{(M+2)/M} ,$$

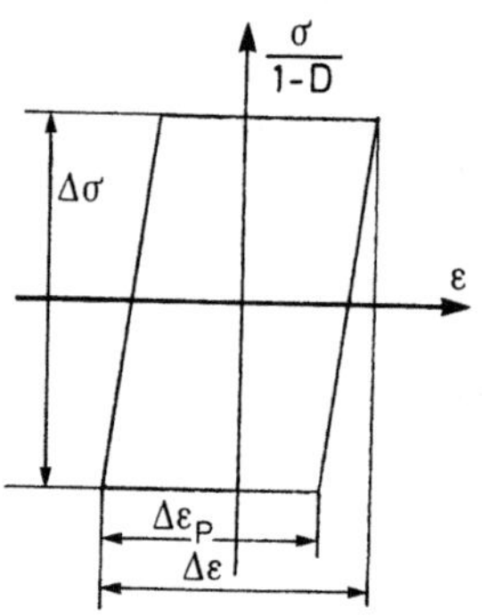

Fig. 3.19. Simplified stress strain cycle

$\Delta\varepsilon_p$ being constant throughout the damage process, the damage is linear with N:

$$D = \int_{N_o}^{N} \frac{\delta D}{\delta N}\delta N = \frac{K_p^2}{4ES}\Delta\varepsilon_p^{(M+2)/M}(N - N_o).$$

The number of cycles to failure is reached when $D = D_c$, with $N_o = \dfrac{p_D}{2\Delta\varepsilon_p}$:

$$D_c = \frac{K_p^2}{4ES}\Delta\varepsilon_p^{(M+2)/M}\left(N_R - \frac{p_D}{2\Delta\varepsilon_p}\right),$$

$$\boxed{N_R = \frac{p_D}{2\Delta\varepsilon_p} + \frac{4ESD_c}{K_p^2}\Delta\varepsilon_p^{-(M+2)/M}.}$$

The power relationship between N_R and $\Delta\varepsilon_p$ is known as the Manson-Coffin law of low cycle fatigue. An example is given in Figure 3.20. For many metallic materials $(M+2)/M$ is of the order of 2.

Coming back to the evolution of D we have

$$D = \frac{K_p^2}{4ES}\Delta\varepsilon_p^{(M+2)/M}(N - N_0)$$

and

$$N_R - N_o = \frac{4ESD_c}{K_p^2}\Delta\varepsilon_p^{-(M+2)/M}.$$

This shows that

$$D = D_c\frac{N - N_o}{N_R - N_o}.$$

In a small range of $\Delta\varepsilon_p$ the critical value of the damage does not vary too much:

$$D_c \simeq \text{const}.$$

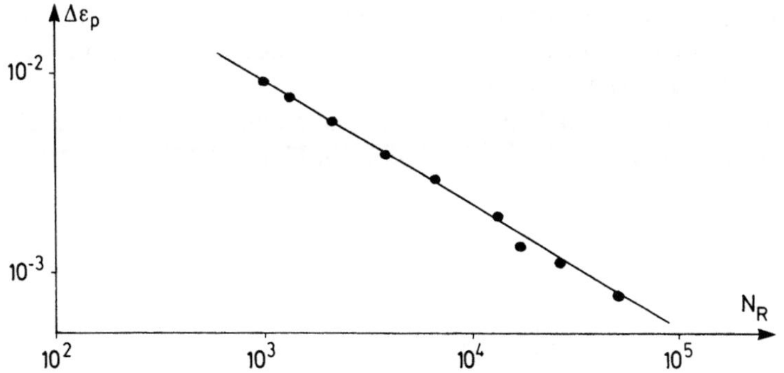

Fig. 3.20. Manson-Coffin curve of low cycle fatigue of a strainhardened steel: AISI 1010 at temperature $T = 20°C$

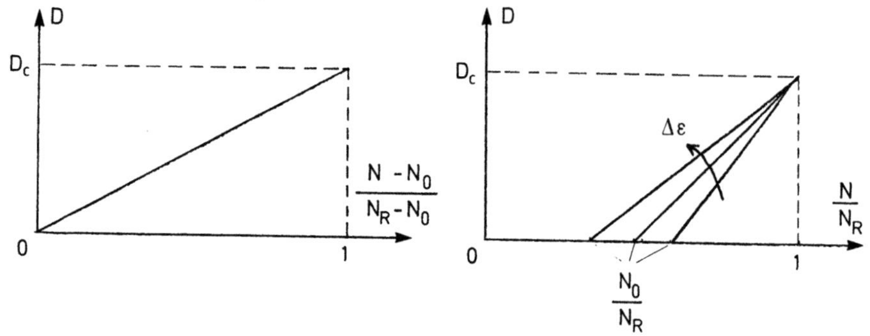

Fig. 3.21. Evolution of the damage as modeled by the simplified theory

Then in practice a single curve exists in the diagram $\left(D, \dfrac{N - N_o}{N_R - N_o}\right)$, but the damage evolution curves differ by N_o in the $\left(D, \dfrac{N}{N_R}\right)$ diagram (Figure 3.21). An example of real low cycle damage evolution measured by the variation of the elasticity modulus is given in Figure 3.22.

Three-dimensional complex loading in time

The kinetic law of damage evolution has to be integrated with the coupled elasto-plastic constitutive equations for the history of the stress or strain under conderation. This can only be done numerically by one of the methods described in Chapter 4. Here again there is a strong influence of the triaxiality.

3.4.2 Exercise on Creep Fatigue Interaction

An important feature of low cycle fatigue of metals at high temperature is the accumulation of fatigue damage and creep damage giving rise to nonlinear effects. This can be modeled by a complete numerical computation of the coupled constitutive equations but the main trend of linear or nonlinear interactions may be pointed out analytically with some approximations.
Consider a periodic one-dimensional stress loading and assume a stabilized cycle in the (stress versus plastic strain) diagram as shown in Figure 3.23.

Calculation of the expression

$$\frac{N_R}{N_R^F} + \frac{N_R}{N_R^c},$$

where:

— N_R is the number of cycles to failure of the creep fatigue process corresponding to some value of D_c;

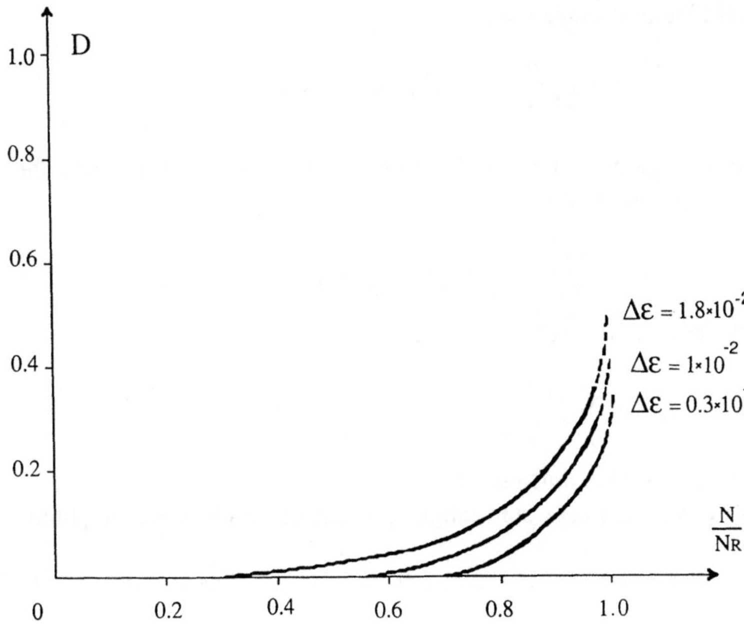

Fig. 3.22. Damage evolution in low cycle fatigue of strain-hardened AISI 1010 steel $10^2 < N_R < 10^4$ cycles

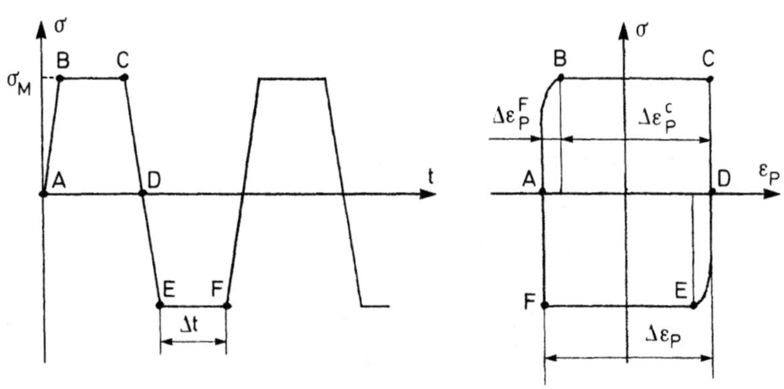

Fig. 3.23. Creep fatigue loading

— N_R^c is the number of cycles to failure of a pure cyclic creep process at constant stress amplitude $\sigma = \sigma_M$ for which the critical value of damage is D_c^c;
— N_R^F is the number of cycles to failure of a pure fatigue process without any hold time,

$\Delta t = 0$, for which $D_c = D_c^F$.
The same damage threshold p_D allows for the above three cases since the same stress amplitude is assumed.

Using the general kinetic damage law,

$$\dot{D} = \frac{\sigma^2}{2ES(1-D)^2}\left|\dot{\varepsilon}_p\right| \text{ if } \varepsilon_p > p_D.$$

1) *Calculate the damage per cycle in the creep fatigue process*, neglecting the variation of damage during that cycle:

$$\frac{\delta D}{\delta N} = \frac{\sigma_M^2}{ES(1-D)^2}\Delta\varepsilon_p,$$

from which the number of cycles to rupture is

$$N_R = \Delta\varepsilon_p^{-1}\left[\frac{p_D}{2} + \frac{ES\left[1-(1-D_c)^3\right]}{3\sigma_M^2}\right].$$

2) *The cycle in Figure 3.23 is partitioned in*
$-(AB+DE)$ considered as a pure fatigue process of plastic strain amplitude $\Delta\varepsilon_p^F$;
$-(BC+EF)$ considered as a pure cyclic creep process of plastic strain amplitude $\Delta\varepsilon_p^c$.
The damage per cycle of the pure fatigue process is

$$\frac{\delta D}{\delta N} = \frac{\sigma_M^2}{ES(1-D)^2}\Delta\varepsilon_p^F,$$

from which the number of cycles to failure is

$$N_R^F = \Delta\varepsilon_p^{F-1}\left[\frac{p_D}{2} + \frac{ES[1-(1-D_c^F)^3]}{3\sigma_M^2}\right].$$

3) *The damage per cycle of the pure cyclic creep process is*

$$\frac{\delta D}{\delta N} = \frac{\sigma_M^2}{ES(1-D)^2}\Delta\varepsilon_p^c$$

and

$$N_R^c = \Delta\varepsilon_p^{c^{-1}}\left[\frac{p_D}{2} + \frac{ES[1-(1-D_c^c)^3]}{3\sigma_M^2}\right].$$

It follows that

$$\frac{N_R}{N_R^F} + \frac{N_R}{N_R^c} = \frac{\Delta\varepsilon_p^F}{\Delta\varepsilon_p}\,\frac{\dfrac{p_D}{2} + \dfrac{ES[1-(1-D_c)^3]}{3\sigma_M^2}}{\dfrac{p_D}{2} + \dfrac{ES[1-(1-D_c^F)^3]}{3\sigma_M^2}} + \frac{\Delta\varepsilon_p^c}{\Delta\varepsilon_p}\,\frac{\dfrac{p_D}{2} + \dfrac{ES[1-(1-D_c)^3]}{3\sigma_M^2}}{\dfrac{p_D}{2} + \dfrac{ES[1-(1-D_c^c)^3]}{3\sigma_M^2}}.$$

The expression $\frac{N_R}{N_R^F} + \frac{N_R}{N_R^c}$ may be smaller or greater than 1 depending upon the relative values of D_c, D_c^F and D_c^c. This property is known as the nonlinear creep fatigue interaction. As $\Delta\varepsilon_p = \Delta\varepsilon_p^F + \Delta\varepsilon_p^c$, the sum of the two ratios is equal to 1 if $D_c = D_c^F = D_c^c$:

$$\frac{N_R}{N_R^F} + \frac{N_R}{N_R^c} = 1,$$

which is the Taira rule of linear creep fatigue interaction.

At very high temperature, the superalloys show a strong nonlinearity in their creep fatigue interaction in the sense that a small amount of creep may considerably reduce the number of cycles to failure as shown in the bottom curve of the interaction diagram in Figure 3.24.

3.4.3 High Cycle Fatigue

If the amplitude of the loading is low, the amplitude of the plastic strain may be very small, even negligible at the mesoscale in comparison with the amplitude of the elastic strain. This corresponds to high values of the number of cycles to failure. For instance,

$$NR > 100000 \text{ cycles.}$$

Another feature which makes the damage analysis of high cycle fatigue difficult is its high degree of localization. Very often only a very small microelement damaged at the free surface of the body gives rise to one microcrack by slips (stage 1) which later propagates perpendicularly to the loading (stage 2). Then, for high cycle fatigue, materials may be considered as quasi-brittle and modeled by a damageable microinclusion embedded in an elastic mesoelement. Its complete numerical analysis, including high cycle fatigue, is performed in Section 4.3. Nevertheless, the

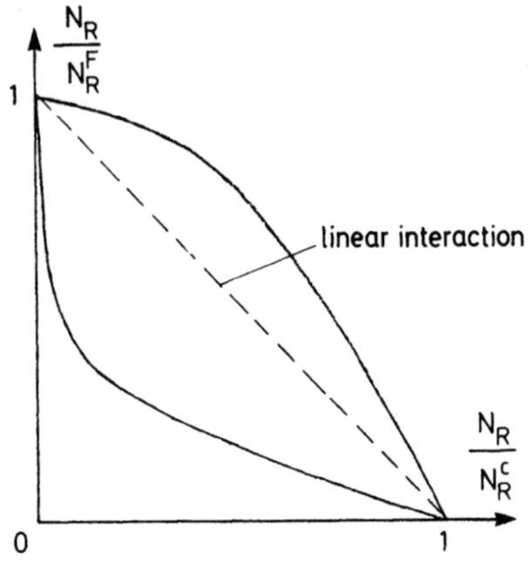

Fig. 3.24. Creep-fatigue interaction diagram

main features of high cycle fatigue may be derived from the simplified analysis of the quasi-brittle damage presented in Section 3.2.2.
The following kinetic law of damage evolution

$$\dot{D} = \frac{\sigma_f^2}{2ES} R_{\nu\left(\frac{\sigma_H}{\sigma_f}\right)} \dot{\varepsilon}_{eq} H(\varepsilon_{eq} - p_D) \text{ with } p_D = \frac{\sigma_u - \sigma_f}{\sigma_f - \dfrac{\sigma_f^2}{\sigma_y}},$$

has been derived from the three hypotheses:

— perfectly plastic behavior of the inclusion with a plastic threshold $\sigma_s^\mu = \sigma_f$, the fatigue limit
— equality of the total strains in the inclusion and in the matrix (Lin-Taylor's hypothesis).
— elastic strain neglected in comparison to the plastic strain in the inclusion,

This law accounts for any kind of three-dimensional high cycle fatigue, ε_{eq} being the accumulated total strain.

One-dimensional case of fatigue

Consider a periodic process of fatigue in tension compression at the mesoscale, stress controlled $\Delta\sigma$ = const. as shown in Figure 3.25. Note that a one-dimensional state of stress at the mesoscale accounts for a three-dimensional state of stress at the microscale due to the mismatch of Poisson's ratio at the mesoscale and the plastic contraction coefficient at the microscale.
In one-dimension at the mesoscale:

$$\frac{\sigma_H}{\sigma_f} = \frac{\sigma}{3\sigma_f}, \quad \dot{\varepsilon}_{eq} = |\dot{\varepsilon}|$$

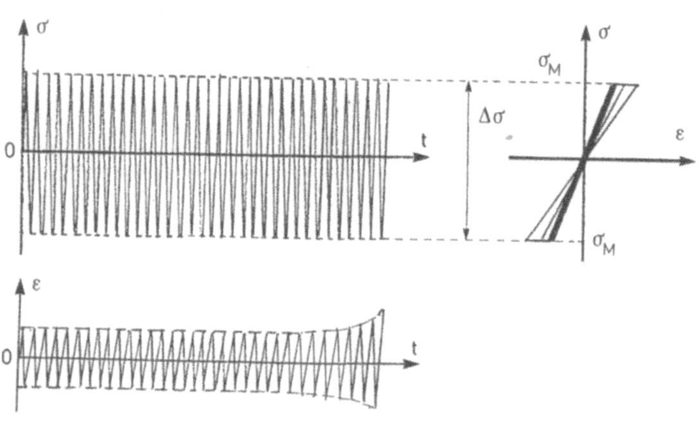

Fig. 3.25. Stress controlled high cycle fatigue

and
$$|\dot{\varepsilon}| = \frac{|\dot{\sigma}|}{E}.$$

Then
$$\dot{D} = \frac{\sigma_f^2}{2E^2 S}\left[\frac{2}{3}(1+\nu)+\frac{1}{3}(1-2\nu)\left(\frac{\sigma}{\sigma_f}\right)^2\right]|\dot{\sigma}|. \text{ if}\begin{cases}\sigma \geq \sigma_f \\ \varepsilon_{eq} \geq p_D\end{cases}.$$

A first integration over one cycle gives the damage per cycle.

$$\frac{\delta D}{\delta N} = \frac{\sigma_f^2}{2E^2 S}4\int_{\sigma_f}^{\frac{\Delta\sigma}{2}}\left[\frac{2}{3}(1+\nu)+\frac{1}{3}(1-2\nu)\left(\frac{\sigma}{\sigma_f}\right)^2\right]d\sigma,$$

$$\frac{\delta D}{\delta N} = \frac{2\sigma_f^2}{E^2 S}\left[(1+\nu)\left(\frac{\Delta\sigma}{2}-\sigma_f\right)+\frac{1-2\nu}{9\sigma_f^2}\left[\left(\frac{\Delta\sigma}{2}\right)^3-\sigma_f^3\right]\right]$$

A second integration as a function of the number of cycles gives the evolution of the damage for the initial condition,

$$N = N_{o(\varepsilon_{eq}=p_D)} \rightarrow D = 0,$$

$$D = \frac{2\sigma_f^2}{E^2 S}\left[(1+\nu)\left(\frac{\Delta\sigma}{2}-\sigma_f\right)+\frac{1-2\nu}{9\sigma_f^2}\left[\left(\frac{\Delta\sigma}{2}\right)^3-\sigma_f^3\right]\right](N-N_o),$$

where N_o is the number of cycles for ε_{eq} to reach the threshold p_D:

$$N_o = \frac{p_D}{2\Delta\varepsilon} = \frac{E p_D}{2\Delta\sigma} = \frac{E\varepsilon_{PD}}{2\Delta\sigma}\frac{\sigma_u-\sigma_f}{\sigma_f-\sigma_f^\mu} = \frac{E\varepsilon_{PD}}{2\Delta\sigma}\frac{\frac{\sigma_u}{\sigma_f}-1}{1-\frac{\sigma_f}{\sigma_y}}.$$

The number of cycles to failure is reached when $D = D_c = D_{1c}\frac{\sigma_u^2}{\tilde{\sigma}^{*2}}$. Taking $D = D_{1c}$ as a "good" approximation,

$$D_{1c} = \frac{2\sigma_f^2}{E^2 S}\left[(1+\nu)\left(\frac{\Delta\sigma}{2}-\sigma_f\right)+\frac{1-2\nu}{9\sigma_f^2}\left[\left(\frac{\Delta\sigma}{2}\right)^3-\sigma_f^3\right]\right](N_R-N_o)$$

or
$$\boxed{N_R = N_o + \frac{D_{1c}}{\frac{2\sigma_f^2}{E^2 S}\left[(1+\nu)\left(\frac{\Delta\sigma}{2}-\sigma_f\right)+\frac{1-2\nu}{9\sigma_f^2}\left[\left(\frac{\Delta\sigma}{2}\right)^3-\sigma_f^3\right]\right]}}.$$

The graph of the amplitude of stress $\Delta\sigma$ as a function of the number of cycles to failure is known as the Woehler curve. An example is given in Figure 3.26. Using the expression for $(N_R - N_o)$, it is possible to express the evolution of the damage in a simpler way:.

$$D = D_{1c}\frac{N-N_o}{N_R-N_o}.$$

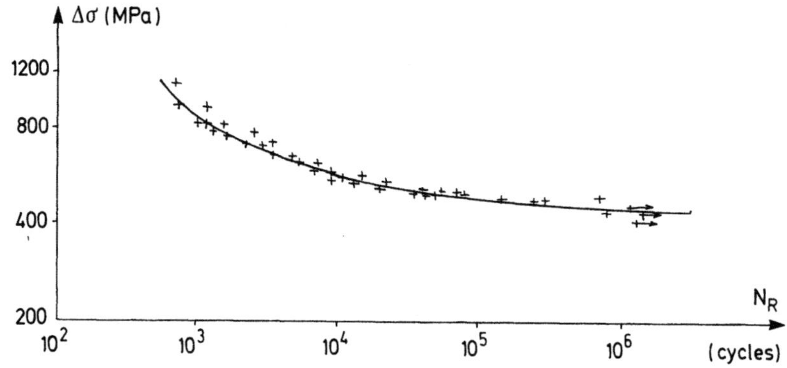

Fig. 3.26. Woehler curve of high cycle fatigue of AISI 316 stainless steel

This shows that the damage is a linear function of the number of cycles. With the approximation $D_c = D_{1c}$, a unit straight line represents the evolution of the damage in the graph $\left(D, \dfrac{N - N_o}{N_R - N_o}\right)$, but as N_o decreases with the amplitude of the stress, different lines represent the evolution of the damage as a function of the number of cycles. These types of damage evolution are shown schematically in Figure 3.27.
An example of real evolution measured by means of elasticity changes is shown in Figure 3.28.

Influence of the mean stress

The previous study has been worked out with tension and compression stresses equal in their absolute values, that is, a mean stress equal to zero. The influence of the mean stress, an important feature of fatigue, can be analyzed numerically as described in Chapter 4, using the concept of unilateral conditions developed in Section 2.3.4, which takes the microcrak closure effect into account. Nevertheless, using a very simple example, let us illustrate how the stress amplitude which induces a certain number of cycles to failure decreases as the mean stress increases.

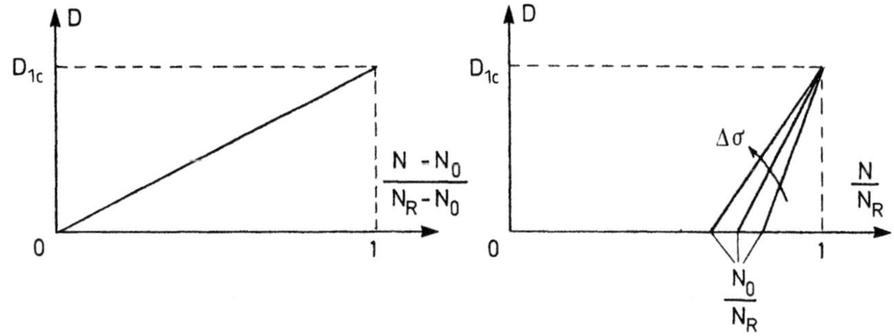

Fig. 3.27. Evolution of high cycle fatigue damage as modeled by the simplified theory

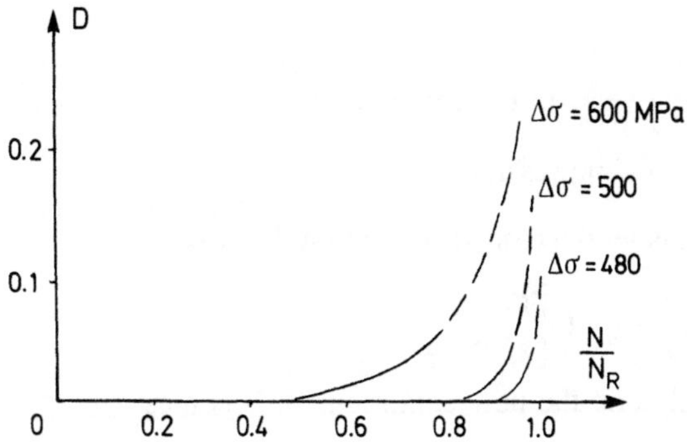

Fig. 3.28. Fatigue damage evolution for AISI 316 stainless steel $10^4 < N_R < 10^6$ cycles

Consider, as above, a periodic process of fatigue loaded in such a way at the mesoscale that we have a one-dimensional loading at the microscale, $R_\nu = 1$. The maximum stress is σ_M, the minimum stress is σ_m and the mean stress is $\overline{\sigma} = \dfrac{\sigma_M + \sigma_m}{2}$, (Figure 3.29). Then, the stress amplitude is

$$\Delta\sigma = \sigma_M - \sigma_m = 2(\sigma_M - \overline{\sigma}).$$

A pure unilateral condition of microcrack closure is assumed for simplicity. From Section 3.1.4.

$$h = 0 \to Y = \frac{1+\nu}{2E} \frac{\langle\sigma_{ij}\rangle\langle\sigma_{ij}\rangle}{(1-D)^2} - \frac{\nu}{2E} \frac{\langle\sigma_{kk}\rangle^2}{(1-D)^2},$$

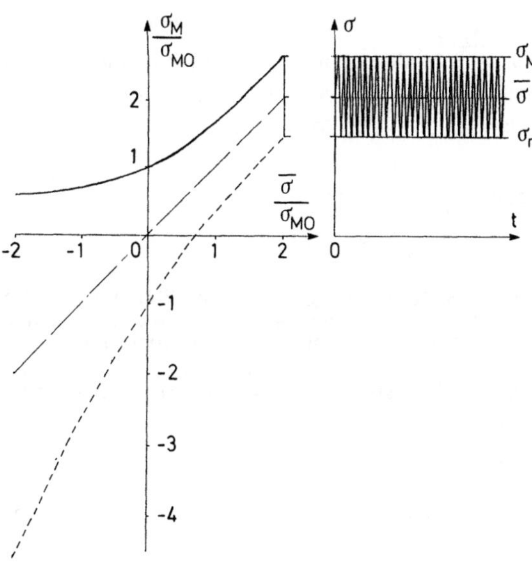

Fig. 3.29. Goodman's diagram (schematic)

which reduces to

$$Y = \frac{\sigma^2}{2E(1-D)^2} \quad \text{in tension } \sigma > 0;$$

$$Y = 0 \quad \text{in compression } \sigma < 0.$$

The kinetic law of quasi-brittle damage developed in Section 3.2.2 was

$$\dot{D} = \frac{Y^\mu}{S}\dot{\varepsilon}_{eq} \quad \text{if} \begin{cases} \sigma_{eq} \geq \sigma_f \\ \varepsilon_{eq} \geq \varepsilon_D \end{cases}.$$

With the results obtained previously and the new conditions, it becomes

$$\dot{D} = \frac{\sigma_f^2}{2E^2 S}\dot{\sigma} \quad \text{in tension}$$

$$\dot{D} = 0 \quad \text{in compression.}$$

The integration of the damage rate concerns only the positive stress taking the microplasticity threshold equal to zero for simplicity

$$\frac{\delta D}{\delta N} = \int_{\sigma_f}^{\sigma_M} \frac{\sigma_f^2}{2E^2 S} d\sigma = \frac{\sigma_f^2}{2E^2 S}(\sigma_M - \sigma_f)$$

and the number of cycles to failure for $D = D_{1c}$ is

$$N_R = N_0 + \frac{D_{1c}}{\dfrac{\sigma_f^2}{2E^2 S}(\sigma_M - \sigma_f)},$$

where N_o remains unchanged, $N_o = \dfrac{Ep_D}{2\Delta\sigma}$ if a non unilateral condition is assumed for the plastic strain. Replacing $\Delta\sigma$ by its value as a function of σ_M and $\overline{\sigma}$ gives the relation which defines all possible combinations of the maximum stress and the mean stress to obtain a given number of cycles to failure N_R:

$$\frac{Ep_D}{4(\sigma_M - \overline{\sigma})} + \frac{D_{1c}}{\dfrac{\sigma_f^2}{2E^2 S}(\sigma_M - \sigma_f)} = N_R$$

If σ_{M0} is the value of the maximum stress σ_M which gives the number of cycles to failure for a zero value of the mean stress, the graph of $\dfrac{\sigma_M}{\sigma_{M0}}$ as a function of $\dfrac{\overline{\sigma}}{\sigma_{M0}}$ is known as Goodman's diagram. Figure 3.29 shows an example corresponding to an hypothetical case where $N_0 = \dfrac{N_R}{2}$:

$$\frac{\overline{\sigma}}{\sigma_{M0}} = \frac{\sigma_M}{\sigma_{M0}} - \frac{1}{2 - \dfrac{\sigma_{M0}}{\sigma_M}}.$$

Experiments show the same trends but the graph is closer to a straight line starting from the point $\sigma_M = \overline{\sigma} = \sigma_u$.

Other parameters may act on the phenomenon of fatigue. The influence of temperature is taken into account by the material parameters which are functions of the temperature. In high cycle fatigue the influence of frequency is negligible; in the model it disappears when using pure plasticity at the microscale and pure elasticity at the mesoscale. The influence of environment can be evaluated by chemical means used to define appropriate internal variables; it is, however beyond, the scope of the damage mechanics.

3.4.4 Exercise on Damage Accumulation

The question of linear or nonlinear accumulation of fatigue damage due to different levels of stress is very important for the applications.

If the damage due to sequences of loading of different amplitudes may be added independently of their order in time up to failure, this property is known as Palmgreen-Miner's rule: If n_i are the numbers of cycles of sequences of periodic loadings defined by there amplitudes $\Delta\sigma_i$, N_R being the number of cycles to failure due to those sequences; and if N_{Ri} is the number of cycles to failure that would exist for the same material submitted to a periodic loading of constant amplitude $\Delta\sigma_i$ up to failure, then

$$N_R = \Sigma n_i,$$

$$\Sigma \frac{n_i}{N_{Ri}} = 1.$$

This "rule" is in fact no longer a rule if a kinetic damage law is used; it is only a consequence of the mathematical properties of the model used. Let us see under which conditions the model developed in Section 3.4.3 has the property of linear or nonlinear accumulation.

Consider two levels of a one-dimensional loading defined in Figure 3.30.

Using the high cycle damage evolution model written as

$$D = D_{1c} \frac{N - N_{o(\Delta\sigma)}}{N_R - N_{o(\Delta\sigma)}},$$

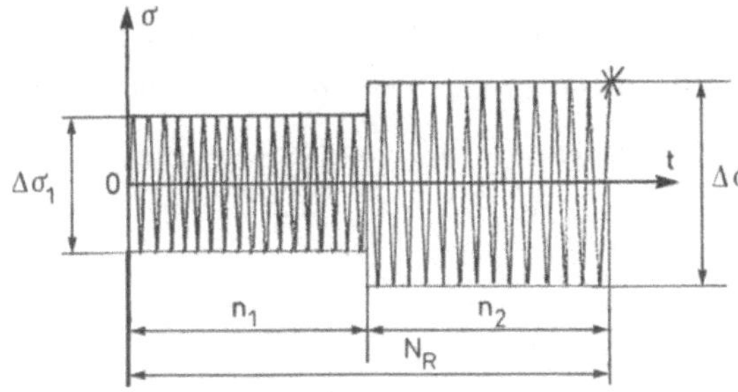

Fig. 3.30. Two-level fatigue case

determine the number of cycles to failure when $n_1 \geq N_{o1}$*:*

$$D_{(n_1)} = D_{1c}\frac{n_1 - N_{01}}{N_{R1} - N_{01}},$$

$$D_{1c(N_R)} = D_{(n_1)} + D_{1c}\frac{N_R - n_1}{N_{R2} - N_{02}},$$

$$N_R = n_1 + (N_{R2} - N_{02})\left(1 - \frac{n_1 - N_{01}}{N_{R1} - N_{01}}\right).$$

Find for which condition the Palmgreen-Miner's rule of linear accumulation would be valid

$$N_R - n_1 = n_2,$$

$$\frac{n_2}{N_{R_2}} = \left(1 - \frac{N_{02}}{N_{R2}}\right)\left(1 - \frac{n_1 - N_{01}}{N_{R1} - N_{01}}\right),$$

$\dfrac{n_1}{N_{R1}} + \dfrac{n_2}{N_{R2}} = 1$ only if $\dfrac{N_{01}}{N_{R1}} = \dfrac{N_{02}}{N_{R2}}$ that is, if a damage threshold proportional to the number of cycles to rupture is assumed.

This corresponds to a unique damage evolution curve independent of the amplitude of stress if D is plotted against N/N_R (see Figure 3.27).

In fact this property is a general property of any differential equation with separated variables. All damage equations of the following general shape show the property of linear accumulation:

$$\dot{D} = f_1(D) f_2(x)\dot{x} \quad \text{with } x = \sigma \text{ or } \varepsilon \text{ or } p$$

if associated with constant initial and final conditions.

In the model studied, the nonlinear accumulation comes from the different values of the damage thresholds.

The experiments show that the Palmgreen-Miner property may be applied if the sequences do not vary too much in amplitude (the variations of p_D may be neglected) but it may give wrong results if the amplitudes differ by more than 10 or 20%.

To quantify these differences please draw the graphs $\dfrac{n_2}{N_{R2}}$ *as a function of* $\dfrac{n_1}{N_{R1}}$ *corresponding to the two-level fatigue sequence for the conditions in Figure 3.31:*

$$\frac{n_2}{N_{R2}} = \left(1 - \frac{n_1}{N_{R1}}\right)\frac{1 - \dfrac{N_{o2}}{N_{R2}}}{1 - \dfrac{N_{o1}}{N_{R1}}}$$

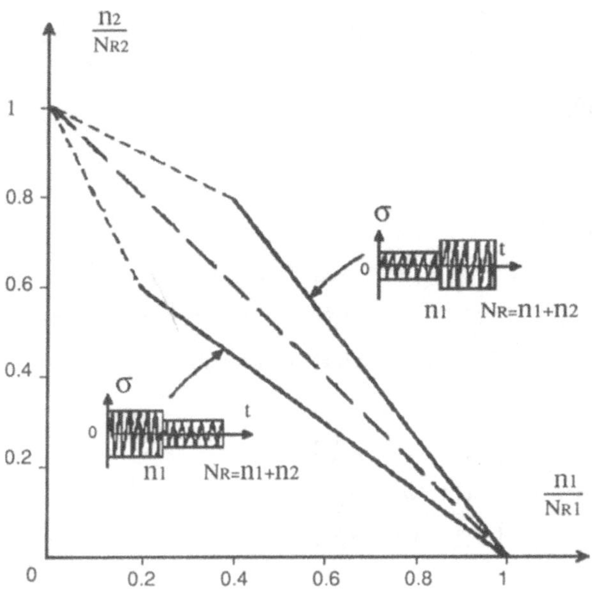

Fig. 3.31. Diagram of nonlinear accumulation of two level fatigue processes:
low level first $N_{o1}/N_{R1} = 0.4$, $N_{o2}/N_{R2} = 0.2$
high-level first $N_{o1}/N_{R1} = 0.2$, $N_{o2}/N_{R2} = 0.4$

3.5 Damage of Interfaces

Composites or multimaterials are made of different layers of possibly different materials bonded together by glue, polymerization or diffusion. One of the main failure mechanisms is the debonding of interfaces. As an interface is "nothing between two somethings!", it is interesting to consider it as a two-dimensional surface medium of zero thickness and to develop a specific damage model in which the inputs are the components of the stresses and the strains which load the interface. These are the components continuous through the interface which may result from a structure calculation of the multimaterial in which the interface is ignored except that the stress vector and the displacement vector are continuous through the interface surface. Then, the calculation of the damage and the debonding acts as a post processor of the structure calculation.

3.5.1 Continuity of the Stress and Strain Vectors

Let us study the basic problem of a mesovolume element made of two different layers A and B bonded together by an interface I of negligible thickness and loaded by a three-dimensional state of stress (Figure 3.32).

On the surface of normal n_3 the equilibrium imposes that

$$\sigma_{i3}^{A} n_3 = \sigma_{i3}^{B} n_3 = \sigma_{i3}^{I} n_3 , \qquad i = 1, 2, 3 .$$

Then, three components of the stress in the interface are known from the stresses in layers A or B:

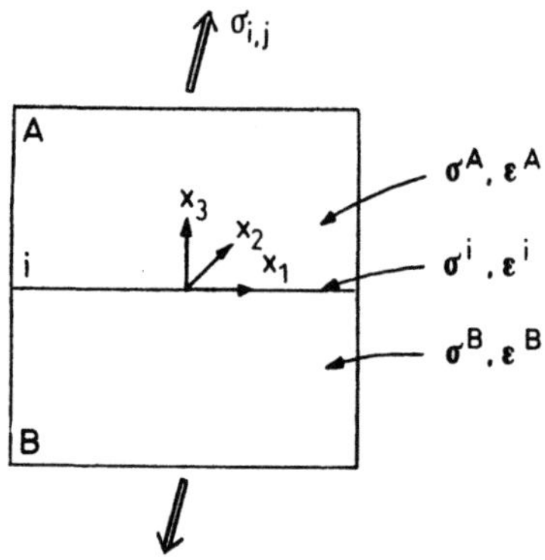

Fig. 3.32. Interface at the mesoscale

$$[\sigma^I] = \begin{bmatrix} \bullet & \bullet & \sigma_{13}^A \\ \bullet & \bullet & \sigma_{23}^A \\ \sigma_{13}^A & \sigma_{23}^A & \sigma_{33}^A \end{bmatrix}.$$

On the interface, there is also continuity of the displacement vector:

$$\forall x_1, x_2 \begin{cases} u_1^A - u_1^B &= 0 \\ u_2^A - u_2^B &= 0 \end{cases} \text{or} \begin{cases} du_1^A(x_1, x_2) &= du_1^B(x_1, x_2) \\ du_2^A(x_1, x_2) &= du_2^B(x_1, x_2) \end{cases}.$$

— From the first equation

$$\frac{\partial u_1^A}{\partial x_1} dx_1 + \frac{\partial u_1^A}{\partial x_2} dx_2 = \frac{\partial u_1^B}{\partial x_1} dx_1 + \frac{\partial u_1^B}{\partial x_2} dx_2,$$

in which $dx_2 = 0 \rightarrow \varepsilon_{11}^A = \varepsilon_{11}^B$,

and $dx_1 = 0 \rightarrow \frac{\partial u_1^A}{\partial x_2} = \frac{\partial u_1^B}{\partial x_2}$.

— From the second equation

$$dx_1 = 0 \rightarrow \varepsilon_{22}^A = \varepsilon_{22}^B,$$

$$dx_2 = 0 \rightarrow \frac{\partial u_2^A}{\partial x_1} = \frac{\partial u_2^B}{\partial x_1}$$

$$\left.\begin{aligned} \frac{\partial u_1^A}{\partial x_2} &= \frac{\partial u_1^B}{\partial x_2} \\ \frac{\partial u_2^A}{\partial x_1} &= \frac{\partial u_2^B}{\partial x_1} \end{aligned}\right\} \begin{aligned} &\therefore \text{addition } \varepsilon_{12}^A = \varepsilon_{12}^B \\ &\because \text{subtraction } \frac{\partial u_1^A}{\partial x_2} - \frac{\partial u_2^A}{\partial x_1} = \frac{\partial u_1^B}{\partial x_2} - \frac{\partial u_2^B}{\partial x_1}. \end{aligned}$$

Then, three components of the strain in the interface, complementary to the known stresses, are known from the strains on layers A or B (plus a rotation condition which does not play any role in classical continuum mechanics):

$$[\varepsilon^I] = \begin{bmatrix} \varepsilon_{11}^A & \varepsilon_{12}^A & \bullet \\ \varepsilon_{12}^A & \varepsilon_{22}^A & \bullet \\ \bullet & \bullet & \bullet \end{bmatrix}.$$

The missing terms in $[\sigma^I]$ and $[\varepsilon^I]$ are of course those which cannot be defined due to the surface character of the interface: no area to support loads in the x_1 and x_2 directions, no length to define strains in the x_3 direction.

3.5.2 Strain Surface Energy Release Rate

The plane strains $\varepsilon_{\alpha\beta}^I$ (with α and $\beta = 1$ or 2) and the antiplane stresses σ_{i3}^I are considered to be independent state and associated variables of the interface. Furthermore, qualitative experiments show that many interfaces exhibit elastic, viscous, plastic and damageable behavior. To follow the State Kinetic Coupling theory used in Chapter 2 it is necessary to split the associated variable $s_{\alpha\beta}^I$ to $\varepsilon_{\alpha\beta}^I$ into an elastic part $s_{\alpha\beta}^e$ and an irreversible part $s_{\alpha\beta}^{an}$

$$s_{\alpha\beta}^I = s_{\alpha\beta}^e + s_{\alpha\beta}^{an},$$

and the same for the state variable e_{i3} to which σ_{i3} is associated

$$e_{i3}^I = e_{i3}^e + e_{i3}^{an}.$$

By analogy with plasticity, a scalar strain hardening state variable π is introduced, its associated variable is named r. The present study is restricted to isothermal behavior which avoid introducing the temperature and the entropy variables. D is the damage variable, the density of microdebondings, $\bar{y}$ is its associated variable as usual.

The state potential is chosen by analogy with the potential used in Chapter 2 except that it is an energy per surface unit;

$$\psi = \frac{1}{2}[k_\varepsilon : \varepsilon^I : \varepsilon^I + k_\sigma : e^e : e^e](1 - D) + \frac{k_\pi \mu}{1 + \mu} \pi^{\frac{1+\mu}{\mu}}$$

(the symbol: means the double contracted tensorial product)
where k_ε, k_σ, k_π and μ are tensors and scalar parameters characteristic of the nature of the interface.
It follows that:

$$s^e = \frac{\partial \psi}{\partial \varepsilon^I}, \quad s_{\alpha\beta}^e \text{ have the dimensions of force per unit length,}$$

$$\sigma^I = \frac{\partial \psi}{\partial e} = k_\sigma : e^e(1 - D), \quad e_{i3}^e \text{ have the dimensions of length,}$$

$$r = \frac{\partial \psi}{\partial \pi} = k_\pi \pi^{1/\mu}, \quad r \text{ has the dimensions of stress,}$$

$$\bar{y} = \frac{\partial \psi}{\partial D} = -\left(\frac{k_\varepsilon}{2} : \varepsilon^I : \varepsilon^I + \frac{k_\sigma}{2} : e : e\right),$$

$y = -\bar{y}$ is the strain surface energy release rate which can be written as a function of the variables known by a classical structure calculation ε^I and σ^I

$$\boxed{y = \frac{k_\varepsilon}{2} : \varepsilon^I : \varepsilon^I + \frac{k_\sigma^{-1}}{2} : \frac{\sigma^I : \sigma^I}{(1-D)^2}} \, .$$

3.5.3 Kinetic Law of Debonding Damage Evolution

The dissipation satisfies the second principle of thermodynamics if

$$s^{an} : \dot{\varepsilon}^I + \sigma^I : \dot{e}^{an} - r\dot{\pi} + y\dot{D} \geq 0 \, .$$

The kinetic law of damage evolution derives from potentials of dissipation which, in the case of stress and strain controlled phenomena, needs to be:

— A viscous-type potential Ω, a nonlinear function of the accumulated equivalent strain ε^I_{eq} defined by

$$\boxed{\dot{\varepsilon}^I_{eq} = \left(\frac{1}{2}\dot{\varepsilon}^I_{\alpha\beta}\,\dot{\varepsilon}^I_{\alpha\beta}\right)^{1/2}}$$

$$\Omega = \frac{y^2}{2}\frac{\varepsilon^{I\eta}_{eq}\,\dot{\varepsilon}^I_{eq}}{S_\varepsilon} \quad \text{if} \quad \varepsilon^I_{eq} > \varepsilon_D \, ,$$

Where η and S_ε are material constants and ε_D, also a material constant, is the damage threshold.

— A plastic-type potential F governed by a criterion f written as a function of an equivalent stress σ^I_{eq} taking into account the unilateral character of interfaces which do not show any damage in compression

$$\boxed{\sigma^I_{eq} = \left\langle \sigma_{33}\,|\sigma_{33}| + \left(\frac{\sigma_R}{\tau_R}\right)^2 (\sigma_{13}^2 + \sigma_{23}^2)\right\rangle^{1/2}} \, .$$

The ratio of the ultimate normal stress σ_R to the ultimate shear stress τ_R models the large difference of strength which often exists between tension and shear in interfaces. $\langle\ \rangle$ are the Macauley brackets, $|\ |$ is the absolute value.

$\sigma^I_{eq} = \sigma_{33}$ in pure tension,

$\sigma^I_{eq} = 0$ in pure compression,

$\sigma^I_{eq} = \dfrac{\sigma_R}{\tau_R}|\sigma_{33}|$ in pure shear,

— a tension decreases the strength in shear,

— a compression increases the strength in shear.

The coupling with damage is neglected in the plastic criterion and potential:

$$f = \sigma_{eq}^{I} - r - \sigma_y = 0\,,$$

$$F = f + \frac{\bar{y}^2}{2S_\sigma}\,,$$

σ_y is a yield stress and S_σ another parameter characteristic of the interface. The kinetic laws of evolution derive from these potentials by the normality rule

$$s_{\alpha\beta}^{an} = \frac{\partial \Omega}{\partial \dot{\varepsilon}_{\alpha\beta}}\,,$$

$$\dot{e}_{i3}^{an} = \frac{\partial F}{\partial \sigma_{i3}} \dot{\lambda}\,,$$

$$\dot{\pi} = -\frac{\partial F}{\partial r} \dot{\lambda} = \dot{\lambda}\,,$$

$$\dot{D} = -\frac{\partial \Omega}{\partial \bar{y}} - \frac{\partial F}{\partial \bar{y}} \dot{\lambda} = y\left(\frac{\varepsilon_{eq}^{I\eta}\, \dot{\varepsilon}_{eq}^{I}}{S_\varepsilon} + \frac{\dot{\lambda}}{S_\sigma}\right),$$

$\dot{\lambda} \geq 0$ is the plastic multiplier determined by the consistency condition $\dot{f} = 0$,

$$\dot{\sigma}_{eq}^{I} - \dot{r} = 0\,,$$

or, together with $r = k_\pi \pi^{1/\mu}$ and $\dot{\pi} = \dot{\lambda}$

$$\dot{\lambda} = \frac{\mu}{k_\pi}\left(\frac{\sigma_{eq}^{I} - \sigma_Y}{k_\pi}\right)^{\mu - 1} \dot{\sigma}_{eq}^{I} \quad \text{if} \begin{cases} f = 0 \\ \dot{f} = 0 \end{cases}.$$

Finally

$$\boxed{\begin{aligned} &\dot{D} = y\left[\frac{\varepsilon_{eq}^{I\eta}}{S_\varepsilon} \dot{\varepsilon}_{eq}^{I} (\text{if } \varepsilon_{eq} \geq \varepsilon_D) + \left(\frac{\sigma_{eq}^{I} - \sigma_Y}{k_\pi}\right)^{\mu - 1} \frac{\mu}{kS_\sigma}\left(\text{if} \begin{cases} f = 0 \\ \dot{f} = 0 \end{cases}\right)\right], \\ &\dot{D} = 0 \ \text{ if not}, \\ &D = D_C \ \rightarrow \ \text{debonding of a Representative Surface Element.} \end{aligned}}$$

The rheological parameters of the interface introduced are k_σ, k_ε, S_ε, η, ε_D, σ_R, τ_R, σ_Y, k_π, μ, D_C: too many! In fact some of them may be identified as group parameters and a simplified model restricts them to four only.

3.5.4 Simplified Model

Different classes of models may be derived from the above expression. Let us restrict ourselves to the two most common mechanisms of debonding:

— debonding in mode I due to the normal stress σ_{33},

— debonding in mode II due to the shear stress σ_{13} or σ_3.

This corresponds to a loading such as $\varepsilon_{\alpha\beta}^{I} \approx 0$ or a material for which k_ε approaches zero ans S_ε is large enough to neglect the first term in the damage law.

Furthermore, consider an isotropic interface which reduces to two the number of independent coefficients in the matrix $[k_\sigma]$, for the same reasons as for isotropic elasticity. Take k_σ for mode I and k'_σ for mode II, then

$$y = \frac{\sigma_{33}^2}{2k_\sigma(1-D)^2} + \frac{\sigma_{13}^2 + \sigma_{23}^2}{k'_\sigma(1-D)^2}.$$

Last assumption to make the model as simple as possible! The yield stress, always difficult to determine, is considered to be zero. Then

$$\dot{D} = y\left(\frac{\sigma_{eq}^I}{k_\pi}\right)^{\mu-1} \frac{\mu}{k_\pi S_\sigma}\,\dot{\sigma}_{eq}^I \quad \text{if} \begin{cases} f = 0 \\ \dot{f} = 0 \end{cases}$$

or

$$\dot{D} = \left[\frac{\sigma_{33}^2}{2k_\sigma(1-D)^2} + \frac{\sigma_{13}^2 + \sigma_{23}^2}{k'_\sigma(1-D)^2}\right]\left(\frac{\sigma_{eq}^I}{k_\pi}\right)^{\mu-1} \frac{\mu}{k_\pi S_\sigma}\langle\dot{\sigma}_{eq}^I\rangle.$$

Qualitative identification

It is possible to relate the parameters k_σ, k'_σ, k_π and S_σ to engineering parameters such as σ_R or τ_R and others easier to measure.

— Consider the case of mode I debonding in pure tension

$$\sigma_{13} = \sigma_{23} = 0, \quad \dot{\sigma}_{33} > 0.$$

The damage law reduces to

$$\dot{D} = \frac{\sigma_{33}^2}{2k_\sigma(1-D)^2}\left(\frac{\sigma_{33}}{k_\pi}\right)^{\mu-1} \frac{\mu}{k_\pi}\frac{\dot{\sigma}_{33}}{S_\sigma}.$$

A mesodebonding occurs for $D = D_C$ when the tensile stress reaches the ultimate tensile stress σ_R. These conditions are obtaned by integration of the damage differential equation with the initial condition of an undamaged state:

$$\sigma_{33} = 0 \rightarrow D = 0$$

$$\int_0^{D_C} (1-D)^2\, dD = \frac{\mu}{2k_\sigma k_\pi^\mu S_\sigma}\int_0^{\sigma_R} \sigma_{33}^{(\mu+1)}\, d\sigma_{33},$$

from which

$$\frac{\mu}{2k_\sigma k_\pi^\mu S_\sigma} = \frac{1-(1-D_C)^3}{\sigma_R^{\mu+2}}\frac{\mu+2}{3}.$$

— Consider now the case of mode II debonding in pure shear:

$$\sigma_{33} = \sigma_{23} = 0, \quad \dot{\sigma}_{13} > 0$$

$$\dot{D} = \frac{\sigma_{13}^2}{k'_\sigma(1-D)^2}\left(\frac{\sigma_R}{\tau_R}\frac{\sigma_{13}}{k_\pi}\right)^{\mu-1} \frac{\mu}{k_\pi}\frac{\sigma_R}{\tau_R}\frac{\dot{\sigma}_{13}}{S_\sigma}.$$

The conditions of a mesodebonding initiation $D = D_C$ when $\sigma_{13} = \tau_R$ are obtained by the same integration as previously and for the same initial conditions:

$$\sigma_{13} = 0 \rightarrow D = 0$$

$$\int_0^{D_C} (1-D)^2\, dD = \frac{\mu}{k'_\sigma k_\pi^\mu S_\sigma} \left(\frac{\sigma_R}{\tau_R}\right)^\mu \int_0^{\sigma_R} \sigma_{13}^{(\mu+1)}\, d\sigma_{13},$$

from which
$$\frac{\mu}{k'_\sigma k_\pi^\mu S_\sigma} = \frac{1-(1-D_C)^3}{\tau_R^{\mu+2}} \frac{\mu+2}{3} \left(\frac{\tau_R}{\sigma_R}\right)^\mu.$$

A more phenomenological writing of the Damage law is obtained by replacing the two calculated group parameters by their values as functions of σ_R, τ_R and D_C:

$$\boxed{\dot{D} = \frac{\mu+2}{3} \frac{1-(1-D_C)^3}{(1-D)^2} \left(\frac{\sigma_{33}^2}{\sigma_R^2} + \frac{\sigma_{13}^2 + \sigma_{23}^2}{\tau_R^2}\right) \left(\frac{\sigma_{eq}^I}{\sigma_R}\right)^{\mu-1} \frac{\langle \dot{\sigma}_{eq}^I \rangle}{\sigma_R}}$$

Quantitative identification

For each nature of interface, four coefficients need to be determined:

σ_R: the ultimate stress in pure tension,
τ_R: the ultimate stress in pure shear,
D_C: the critical damage at debonding,
μ: the exponent characteristic of the nonlinearity

Even in the case of glue the identification tests must be performed on a multimaterial that is the interface and its two adjacent layers because an isolated interface, glue for example, has different properties than a built-in interface.

— A pure tensile test allows the determination of σ_R.
— A pure shear test with unloading allows the determination of τ_R and D_C by extrapolation of the Damage evolution measured by

$$D = 1 - \frac{\tilde{E}}{E} = 1 - \frac{\tilde{G}}{G},$$

$\tilde{G}$ being the damaged shear modulus and G its value in the undamaged state. This formula using $G = E/2(1+\nu)$ assumes that the Poisson's ratio is not affected by the damage which is exact in the case of isotropic damage.

— The best way to obtain the exponent μ is to perform fatigue tests in tension ($\sigma_M = 0$, σ_M) or in shear and to deduce μ from the slope of the Woehler curve giving the logarithm of the stress amplitude as a function of the logarithm of the number of cycles to debonding.

The equation of the Woehler curve is obtained by integration of the above damage law over one cycle first neglecting the variation of D:

$$\frac{\partial D}{\partial N} = \frac{\mu+2}{3} \frac{1-(1-D_C)^3}{(1-D)^2} \int_0^{\sigma_M} \left(\frac{\sigma_{33}}{\sigma_R}\right)^{\mu+1} \frac{d\sigma_{33}}{\sigma_R},$$

$$\frac{\partial D}{\partial N} = \frac{1}{3} \frac{1-(1-D_C)^3}{(1-D)^2} \left(\frac{\sigma_M}{\sigma_R}\right)^{\mu+2},$$

and then as a function of the number of cycles with the conditions

$$N = 0 \rightarrow D = 0,$$

$$N = N_R \leftarrow D = D_C,$$

$$\int_0^{D_C} \frac{3(1-D)^2}{1-(1-D_C)^3} \delta D = \int_0^{N_R} \left(\frac{\sigma_M}{\sigma_R}\right)^{\mu+2} \delta N,$$

$$N_R = \left(\frac{\sigma_R}{\sigma_M}\right)^{\mu+2}.$$

Having obtained the Woehler curve from experiments we calculate the exponent μ from

$$\mu = \frac{\operatorname{Log} N_R}{\operatorname{Log} \sigma_R - \operatorname{Log} \sigma_M} - 2.$$

3.5.5 Exercise on a Debonding Criterion for Interfaces

A simplified criterion for debonding of interfaces loaded in proportional loading consists in writing

$$\sigma_{eq}^I = \sigma_R,$$

or

$$\boxed{\left\langle \frac{\sigma_{33}|\sigma_{33}|}{\sigma_R^2} + \frac{\sigma_{13}^2 + \sigma_{23}^2}{\tau_R^2} \right\rangle = 1.}$$

— Show that this criterion is the exact solution given by the damage law when integrated in the case of proportional loading in shear and tension:

$$\sigma_{33} > 0 \rightarrow \frac{\sigma_{33}^2}{\sigma_R^2} + \frac{\sigma_{13}^2 + \sigma_{23}^2}{\tau_R^2} = \left(\frac{\sigma_{eq}^I}{\sigma_R}\right)^2.$$

The integration reduces to

$$\int_0^{D_C} \frac{3(1-D)^2}{1-(1-D_C)^3} dD = (\mu+2) \int_0^{\sigma_{eq}^I} \left(\frac{\sigma_{eq}^I}{\sigma_R}\right)^{\mu+1} \frac{d\sigma_{eq}^I}{\sigma_R},$$

$$1 = \left(\frac{\sigma_{eq}^I}{\sigma_R}\right)^{\mu+2},$$

$$\sigma_{eq}^I = \sigma_R.$$

— In the case of shear and compression, the integration is more complex due to the sign of $\sigma_{33} < 0$ but the numerical values are not far from those given by $\sigma_{eq}^I = \sigma_R$.

For example show that for a constant value of $\sigma_{33}/\sigma_R = -1$, $\sigma_{23} = 0$ and $\mu = 4$ the integration of the damage law gives $\sigma_{13}/\tau_R = 1.371$ and the simple criterion $\sigma_{13}/\tau_R = 1.414$.

Damage law:

$$dD = \frac{6}{3}\frac{1-(1-D_C)^3}{(1-D)^2}\left(1+\frac{\sigma_{13}^2}{\tau_R^2}\right)\left(-1+\frac{\sigma_{13}^2}{\tau_R^2}\right)^{3/2}\left(-1+\frac{\sigma_{13}^2}{\tau_R^2}\right)^{-1/2}\frac{\sigma_{13}}{\tau_R^2}\,d\sigma_{13}\,.$$

Integration:

$$1 = 6\int_0^{\sigma_{13}}\left(1+\frac{\sigma_{13}^2}{\tau_R^2}\right)\left(-1+\frac{\sigma_{13}^2}{\tau_R^2}\right)\frac{\sigma_{13}}{\tau_R^2}\,d\sigma_{13}$$

$$\frac{\sigma_{13}^6}{\tau_R^6} - \frac{3\sigma_{13}^2}{\tau_R^2} = 1 \;\rightarrow\; \frac{\sigma_{13}}{\tau_R} = 1.375\,.$$

Criterion:

$$-1+\frac{\sigma_{13}}{\tau_R^2} = 1 \;\rightarrow\; \frac{\sigma_{13}}{\tau_R} = \pm\sqrt{2}\,.$$

— Graph of the debonding criterion in the tension-shear plane (Figure 3.33):

$$\frac{\sigma_{33}\,|\sigma_{33}|}{\sigma_R^2} + \frac{\sigma_{13}}{\tau_R^2} = 1\,,$$

$\sigma_{33} > 0 \rightarrow$ half a circle,

$\sigma_{33} < 0 \rightarrow$ two branches of an hyperbola.

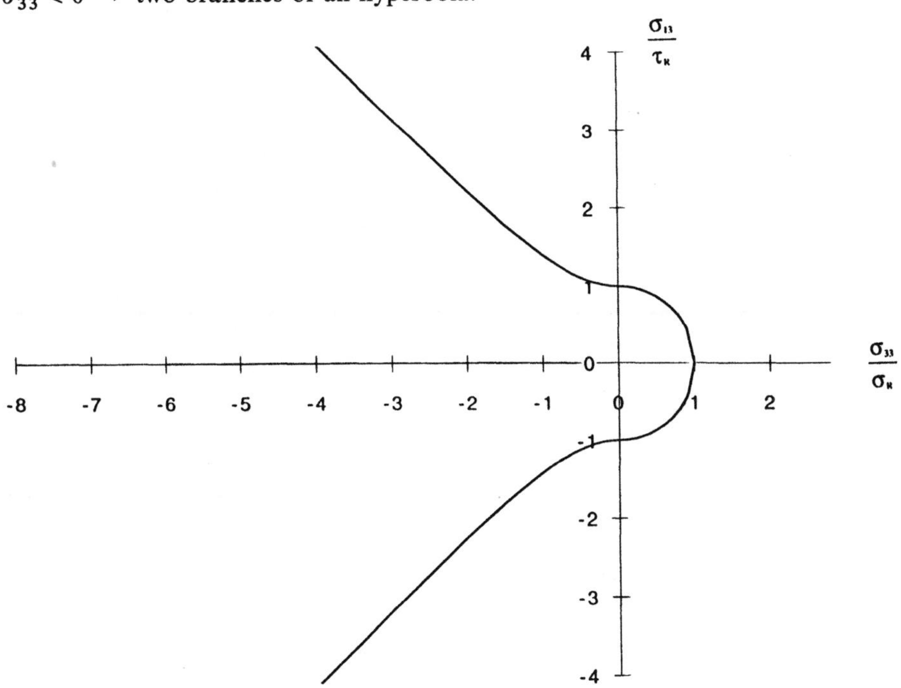

Fig. 3.33. Debonding criterion (from J.P. Sermage)

3.6 Table of material parameters

Users should be aware that all numbers of the following table are to be considered as orders of magnitude only, because each set corresponds to a specific material which may show slightly different properties from another one identically denominated. When performing a structural calculation, use available characteristics chosen with the name of the material and be prepared to perform calculation of corrections with the numbers identified, if possible in situ, on the real material of the real structure. Furthermore, keep in mind that each parameter is related to a specific model. For example, the value of the yield stress of a given material may be different for a perfectly plastic model, a linear kinematic hardening model or an isotropic hardening model.
Dear ''identify-or'', each time you perform an identification please write the results in the following chart and send it to me! Thank you.

Material	Temps = °C	Elasticity		Plasticity			Strain hardening				Viscoplasticity		DAMAGE			Crack closure
		E MPa	ν	σ_f MPa	σ_y MPa	σ_u MPa	X_∞ MPa	γ	R_∞ MPa	b	K_∞ MPa	n s^{-1}	S MPa	ε_{pD}	D_{1c}	h
Steel AISI 1010	20	190 000	0.28	200	320	700					–	–	2.4	0.44	0.2	0.2
Stainless steel AISI 316	20	200 000	0.30	180	260	700	200	2	300	1	–	–	7	0.10	0.15	0.2
	600	140 000	0.32	3	6	–					150	12	0.2	0	0.5	0.5
Copper	20	100 000	0.33	100	190	300					–	–	0.4	0.35	0.85	0.6
Aluminium alloy 2024	20	72 000	0.32	250	300	500					–	–	1.7	0.03	0.23	0.2
Ceramic Alumina at micro scale	20	400 000	0.2	300	306	500	–	–	–	–	–	–	6	0.10	0.99	0.2
Concrete in tension at micro scale	20	30 000	0.2	1.5	2.5	3	–	–	–	–	–	–	$2.5 10^{-7}$	0	0.2	0.2

Chapter 4

Analysis of Crack Initiation in Structures

The fundamental reason for studing damage mechanics is to understand why and how materials break. Together with physics, metallurgy and chemistry this knowledge allows us to improve the mechanical properties of materials and to design new multimaterials. The practical reason for studing damage mechanics is to predict when materials, as they are currently made, will break upon submission to mechanical and thermal loading. This involves the analysis of real components of structures in real or presumed situations.

This analysis requires two main steps: modeling behavior of the material and solving the boundary value problem of stress, strain and damage fields in the structure. The first step has been studied in detail in the previous three chapters (the constitutive equations for plasticity and viscoplasticity are given in Section 2.1.4; the constitutive equations for damage are given in Chapter 3). In the present chapter we shall deal with the second step, by considering how these constitutive equations may be integrated in the classical framework of continuum mechanics, the basic variables of which are stress, strain, temperature and time. In the first section, we review the classical equations and methods. Then, several levels of computation are developed, whereby the coupling between damage and strain is: neglected for a rough approximation; taken into account only for the most damaged points in small-scale damage zones; or, taken into account everywhere for a fully coupled analysis of structure in large-scale damaged zones.

4.1 Stress-Strain Analysis

Consider the general problem of a structure S defined by its geometrical contour ∂S on which a load density $\vec{F}_{(t)}$ given on ∂S_F and displacements $\vec{u}_{(t)}$ given on $\partial S_u(\partial S_F + \partial S_u = \partial S)$ are applied as a function of'time t. A body force density $\vec{f}_{(t)}$ may be applied in S and the temperature distribution in S is given by $T_{(M,t)}$ (Figure 4.1).

The mechanical properties of the material are represented by the strain and damage constitutive equations possible with initial values of stress (residual stresses), of strain hardening (components obtained by metal forming, for example) and of damage (analysis of the residual life of an accidentally damaged component, for example).

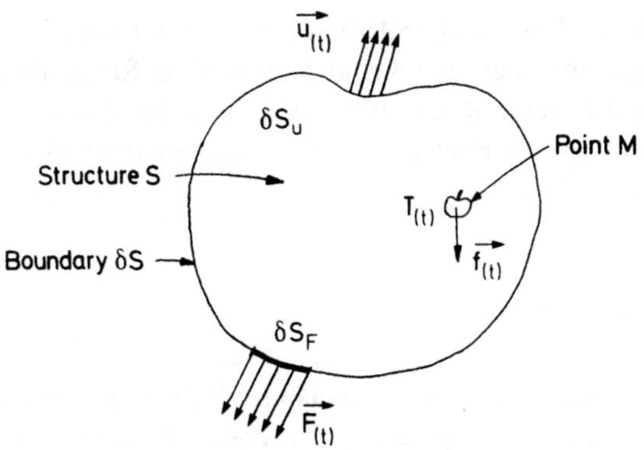

Fig. 4.1. Boundary value problem

The general equations of the problem are:

— The equilibrium equations derived from the principle of virtual work,

$$\sigma_{ij,j} + f_i = \rho \ddot{u}_i \quad \text{in } \mathbf{S}$$

ρ is the density and $\ddot{\vec{u}} = \dfrac{d^2\vec{u}}{dt^2}$ is the acceleration

— The strain-displacement equations, given here for the small deformation theory:

$$\varepsilon_{ij} = \frac{1}{2}(u_{i,j} + u_{i,j}) \quad \text{in } \mathbf{S}$$

— The set of constitutive equations, given in Sections 2.1.4 and 3.1:

$$\varepsilon_{ij} = \varepsilon^e_{ij} + \varepsilon^p_{ij},$$
$$\varepsilon^e_{ij} = \varepsilon^e_{ij}(\sigma_{ij}, D, T),$$
$$\dot{\varepsilon}^p_{ij} = \dot{\varepsilon}^p_{ij}(\dot{\sigma}_{ij}, \sigma_{ij}, X_{ij}, R, D, T),$$
$$\dot{R} = \dot{R}(\dot{\sigma}_{ij}, \sigma_{ij}, X_{ij}, R, D, T),$$
$$\dot{X}_{ij} = \dot{X}_{ij}(\dot{\sigma}_{ij}, \sigma_{ij}, R, D, T),$$
$$\dot{D} = \dot{D}(\sigma_{ij}, D, \varepsilon^p_{ij}, T).$$

— The boundary conditions

$$\sigma_{ij} n_j = F_i \text{ given on } \partial S_F,$$
$$u_i = u_i \text{ given on } \partial S_u,$$
$$T = T \text{ given on } S + \partial S.$$

— The initial conditions

$$\sigma_{ij}(t = 0) = \sigma^0_{ij}$$
$$R(t = 0) = R_0,$$
$$X_{ij}(t = 0) = X^0_{ij},$$
$$D(t = 0) = D_0.$$

Due to the complexity of the partial differential equations and the nonlinearities introduced by the constitutive equations, there is no analytical method for solving this type of problem, in the general case. It can only be solved by means of approximations in some cases and by numerical timelike incremental analysis, usually with the aid of computers.

4.1.1 Stress Concentrations

Regarding damage analysis in structures, the critical points at which a mesocrack may be initiated are almost always located in regions of stress concentration caused by sharp variation of geometry: holes, notches, etc. This means that a high degree of accuracy of stress strain analysis is needed only in those regions.

There is no general method for easily calculating stress concentrations, not even for cases of elasticity without any plastic strain or damage. A useful concept is the stress concentration factor K_T, defined, at one point loaded in one dimension, as the ratio of the stress σ^A at the maximum loaded point A to the nominal stress that would exist in the absence of any effect of the geometrical perturbation on the stress field, σ_A^0:

$$\boxed{K_T = \frac{\sigma^A}{\sigma_A^0}.}$$

Figure 4.2 gives an example of a small circular hole in a large elastic plate. On the boundary of the hole where $x_2 = 0$ the stress is one-dimensional.

Many particular geometries have been analyzed either by approximate analytical methods, photoelasticity or the numerical finite element method. Some classical results taken from handbooks listed in the literature references are given below. All concern isotropic elasticity.

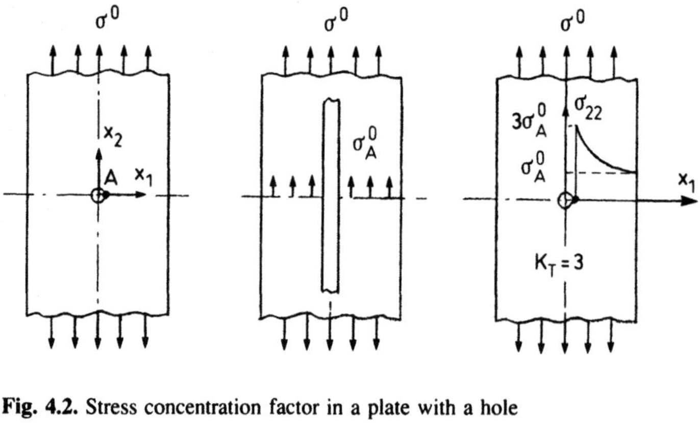

Fig. 4.2. Stress concentration factor in a plate with a hole

—**Elliptic holes in infinite plates under tension** (Figure 4.3)

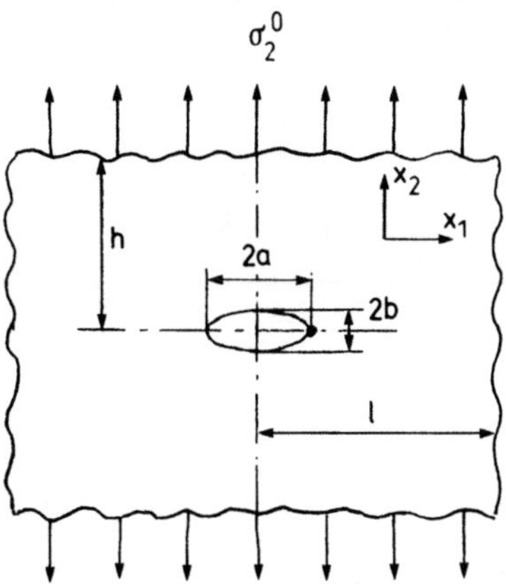

Fig. 4.3.

"Infinite" plate $\rightarrow \begin{cases} l/a & > 10 \\ h/b & > 10 \end{cases}$

Stress concentration factor at A and in the direction of σ_2^0

$$K_T = \frac{\sigma_{22}^0}{\sigma_2^0} = 1 + 2\frac{a}{b}$$

Examples: —circular hole $\frac{a}{b} = 1 \rightarrow K_T = 3$

—horizontal sharp ellipse $\frac{a}{b} = 10 \rightarrow K_T = 21$

—vertical sharp ellipse $\frac{a}{b} = \frac{1}{10} \rightarrow K_T = 1.2$

—**Circular holes in infinite plates under biaxial tension** (Figure 4.4)

$$K_T = \frac{\sigma_{22}^A}{\sigma_2^0} = 3 - \frac{\sigma_1^0}{\sigma_2^0}$$

valid for $-1 \leq \frac{\sigma_1}{\sigma_2} \leq 1$

Examples: —equibiaxial tension

$$\frac{\sigma_1}{\sigma_2} = 1 \rightarrow K_T = 2$$

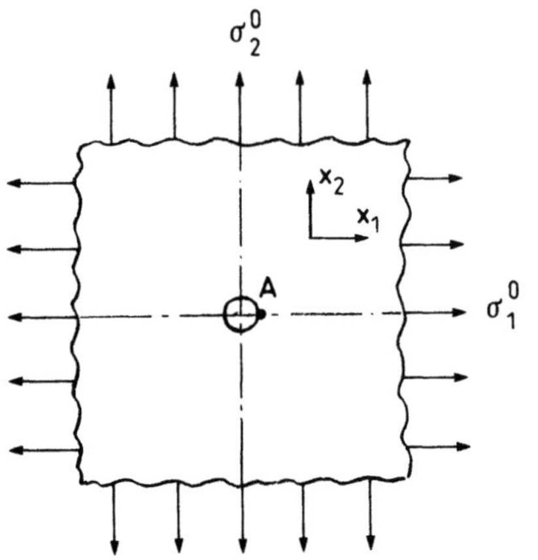

Fig. 4.4.

—equal tension and compression

$$\frac{\sigma_1}{\sigma_2} = -1 \rightarrow K_T = 4$$

—Edge effect in circular holes in plates under tension (Figure 4.5)

$$K_T = \frac{\sigma_{22}^{A \text{ or } B \text{ or } C}}{\sigma_2^0}$$

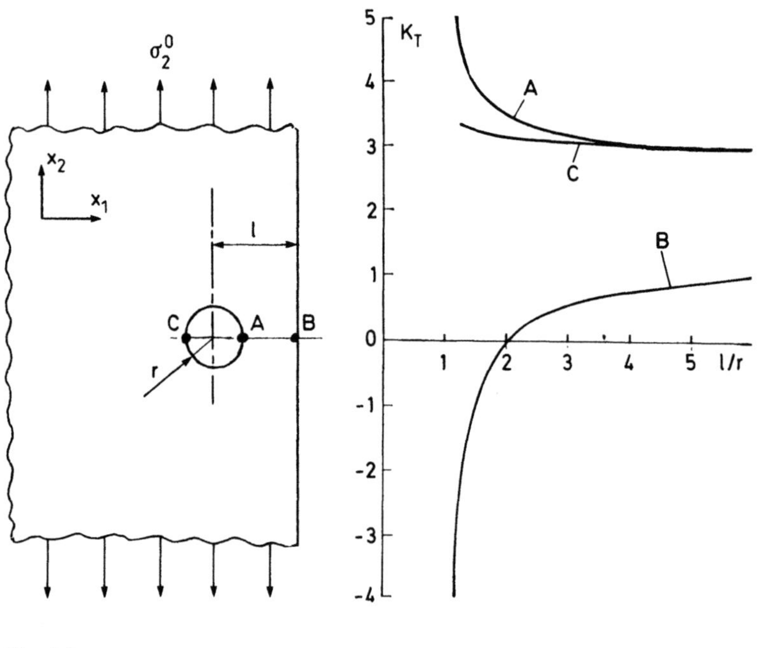

Fig. 4.5.

Example: $\frac{1}{r} = 2 \rightarrow \begin{cases} K_T^A & = 3.5 \\ K_T^B & = 0 \\ K_T^C & = 3.15 \end{cases}$

– **Multihole plates under biaxial tension** (Figure 4.6)

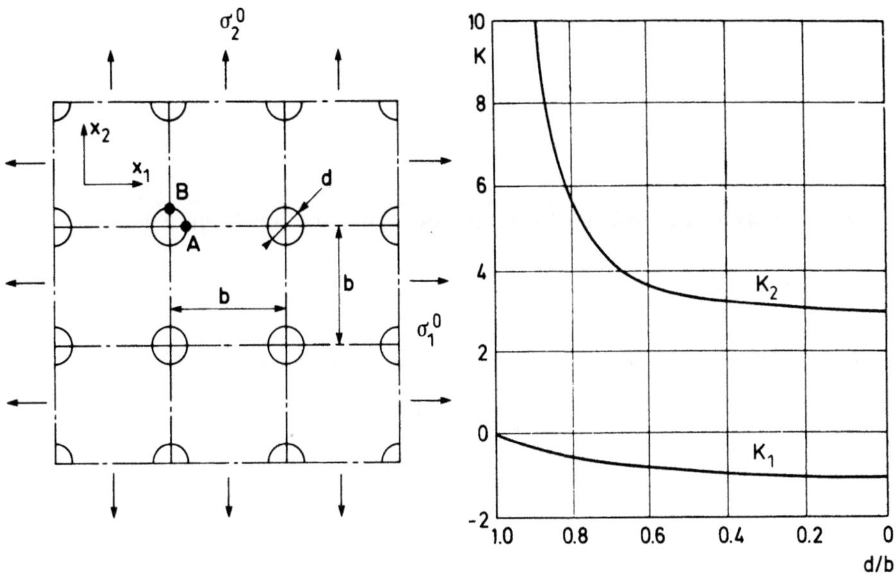

Fig. 4.6.

$$\sigma_{22}^A = K_1\sigma_2^0 + K_2\sigma_1^0$$

$$\sigma_{22}^B = K_2\sigma_2^0 + K_1\sigma_1^0$$

Example: $d/b = 0.4 \quad \sigma_{22}^A = 3.3\sigma_2^0 - 1\sigma_1^0$

– **Two V notches in members of rectangular section** (Figure 4.7)

$K_r = \frac{\sigma_{22}^A}{\sigma_{22A}^0}$

r/b	0.05	0.10	0.15	0.20	0.25	0.30	0.40	0.50
Tension	4.1	3.0	2.5	2.2	2.0	1.9	1.7	1.55
Bending	2.9	2.2	1.9	1.7	1.6	1.5	1.4	1.3

$$\sigma_{22A}^0 = \sigma_2^0 \frac{l}{b}$$

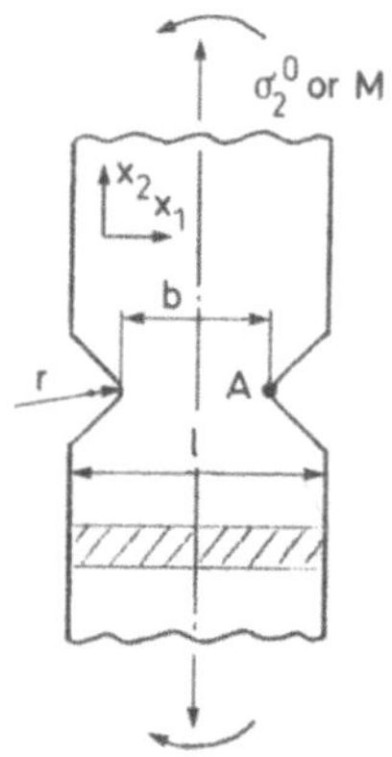

Fig. 4.7.

—One V notch in member of rectangular cross section (Figure 4.8)

Approximation for $h/b \ll 1$

$$\text{Tension} \rightarrow K_T = \frac{\sigma_{22}^A}{\sigma_{22A}^0} = 1 + 2\sqrt{h/r}, \quad \sigma_{22A}^0 = \sigma_2^0 \frac{b+h}{b}$$

$$\text{Bending} \rightarrow K_T = \frac{\sigma_{22}^A}{\sigma_{22A}^0} = 1 + 2\sqrt{h/r}$$

Example: $h/r = 4 \rightarrow K_T = 5$

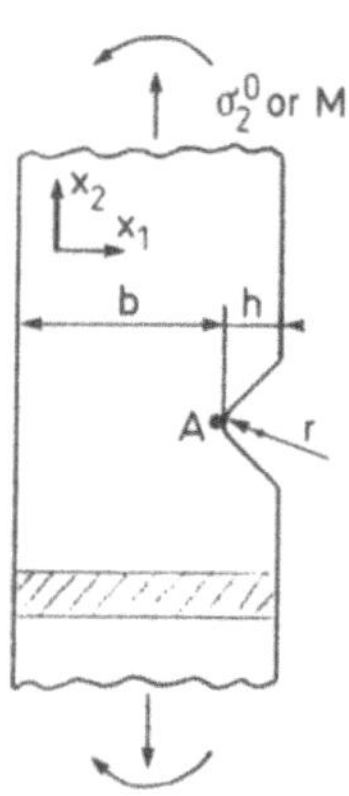

Fig. 4.8.

—Square shoulder with fillet in rectangular bar (Figure 4.9)

$K_T = \dfrac{\sigma_{22}^A}{\sigma_{22A}^0}$

	h/r \ r/b	0.05	0.10	0.20	0.27	0.50	1.0
Tension	0.5	1.70	1.60	1.53	1.47	1.39	1.21
	1.0	1.93	1.78	1.67	1.59	1.42	1.22
	1.5	...	1.89	1.72	1.65	1.43	1.23
	2.0	...	1.95	1.80	1.70	1.44	1.23
	3.5	...	2.10	1.93	1.78	1.47	1.24

$K_T = \dfrac{\sigma_{22}^A}{\sigma_{22A}^0}$

	h/r \ r/b	0.05	0.10	0.20	0.27	0.50	1.0
	0.5	1.61	1.49	1.39	1.34	1.22	1.07
	1.0	1.91	1.70	1.48	1.38	1.22	1.08
Bending	1.5	2.00	1.73	1.50	1.39	1.23	1.08
	2.0	...	1.74	1.52	1.39	1.23	1.09
	3.5	...	1.76	1.54	1.40	1.23	1.10

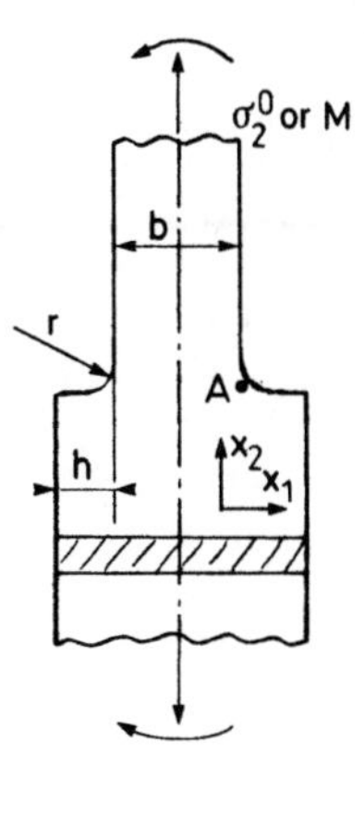

Fig. 4.9.

– **Square shoulder with fillet in circular shaft subjected to torsion** (Figure 4.10)

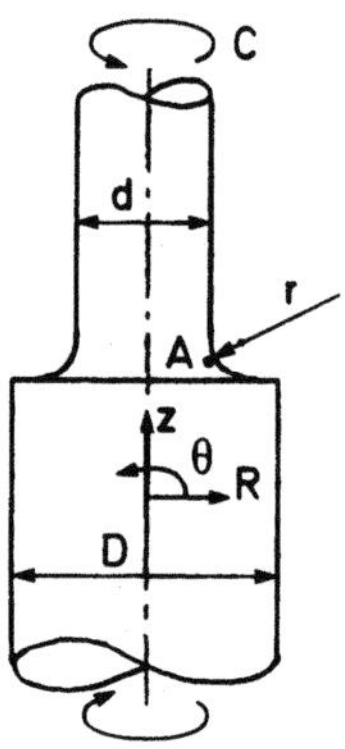

Fig. 4.10.

$K_T = \frac{\sigma_{R\theta}^A}{\sigma_{R\theta A}^0}$

	D/d ↘ r/d	0.005	0.01	0.02	0.03	0.04	1.06	0.08	0.10	0.12
Bending	2.00	...	3.0	2.25	2.00	1.82	1.65	1.51	1.44	1.39
	1.33	...	2.7	1.16	1.91	1.76	1.60	1.48	1.40	1.35
	1.20	3.00	2.5	2.00	1.75	1.62	1.50	1.40	1.34	1.30
	1.09	2.20	1.88	1.53	1.40	1.30	1.20	1.16	1.15	1.15

– V **notch in circular shaft subjected to torsion** (Figure 4.11)

$K_T = \frac{\sigma_{R\theta}^A}{\sigma_{R\theta}^0}$

α ↘ h/r	0.5	1	3	5	9
0°	1.85	2.01	2.66	3.23	4.54
60°	1.84	2.00	2.54	3.06	3.90
90°	1.81	1.95	2.40	2.64	3.12
120°	1.66	1.75	1.95	2.06	2.13

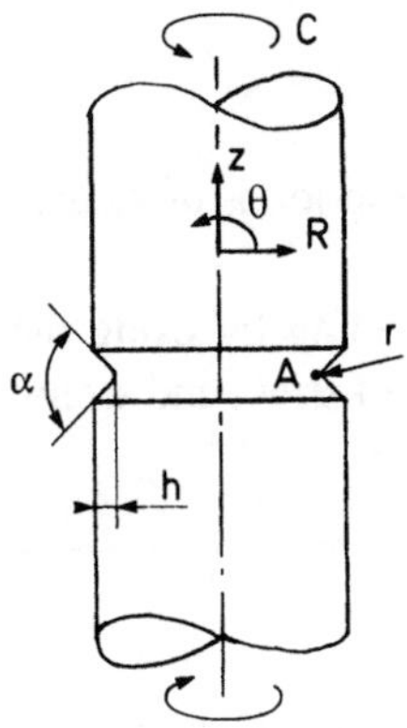

Fig. 4.11.

4.1.2 Neuber's Method (Neuber 1947)

When plasticity occurs, even locally, it is much more complicated. Neuber's method allows us to calculate the elastoplastic stress concentration from a purely elastic calculation. It applies for all notch problems at the root of which there is a stress concentration.

The basic heuristic hypothesis, checked for thousands of applications, postulates that for a one-dimentional elastic problem, the product of the stress by the strain $\sigma \cdot \varepsilon$ at the most loaded point is equal to the same product for the same problem analyzed in pure elasticity $\sigma_E \cdot \varepsilon_E$

$$\boxed{\sigma \cdot \varepsilon = \sigma_E \cdot \sigma_E = \frac{\sigma_E^2}{E}}$$

Assuming the elastic problem to be solved, σ_E and ε_E are known. The second equation with which to calculate the two variable σ and ε is the elastoplastic constitutive equation. For the simple case of monotonic loading, let us take some stress function representing the stress strain curve:

$$\varepsilon = g_\varepsilon(\sigma).$$

Then

$$\sigma \cdot g_\varepsilon(\sigma) = \frac{\sigma_E^2}{E} = \text{const}.$$

is the equation with which to find the stress at the root of an elasto plastic notch. A simple geometrical construction follows from the above two equations.

In the stress strain coordinate system;

$$\sigma \cdot \varepsilon = \frac{\sigma_E^2}{E}$$

is an hyperbola containing the point $\left(\sigma = \sigma_E, \varepsilon = \frac{\sigma_E}{E}\right)$.

The other equation,

$$\sigma = g_{\varepsilon}^{-1}(\varepsilon),$$

is the classical elastic strain hardening stress strain curve. The solution is obviously given by the intersection of these two curves (Figure 4.12).

This method may be generalized straighforwardly to cyclic loading by using the cyclic elasto plastic stress strain curve g_c to relate the amplitude of stress and strain:

$$\Delta\sigma \cdot \Delta\varepsilon = \Delta\sigma_E \cdot \Delta\varepsilon_E = \frac{(\Delta\sigma_E)^2}{E},$$

$$\Delta\varepsilon = g_c(\Delta\sigma).$$

The generalization to a three-dimensional state of stress is also easy if we restrict the application to proportional loadings.

The basic heuristic hypothesis is

$$\sigma_{ij}\varepsilon_{ij} = \sigma_{ij}^E\varepsilon_{ij}^E.$$

With the same notations as previously,

$$\sigma_{ij}\varepsilon_{ij} = \sigma_{ij}\left(\varepsilon_{ij}^e + \varepsilon_{ij}^p\right).$$

$^*\sigma_{ij}\varepsilon_{ij}^e = 2w_e$, the elastic strain energy which has been calculated in Section 2.1.2 as:

$$2w_e = \frac{2(1+\nu)}{3E}\sigma_{eq}^2 + \frac{3(1-2\nu)}{E}\sigma_H^2.$$

with $D = 0$,

$$\sigma_{ij}\varepsilon_{ij}^p = \left(\sigma_{ij}^D + \sigma_H\delta_{ij}\right)\varepsilon_{ij}^p = \sigma_{ij}^D\varepsilon_{ij}^p,$$

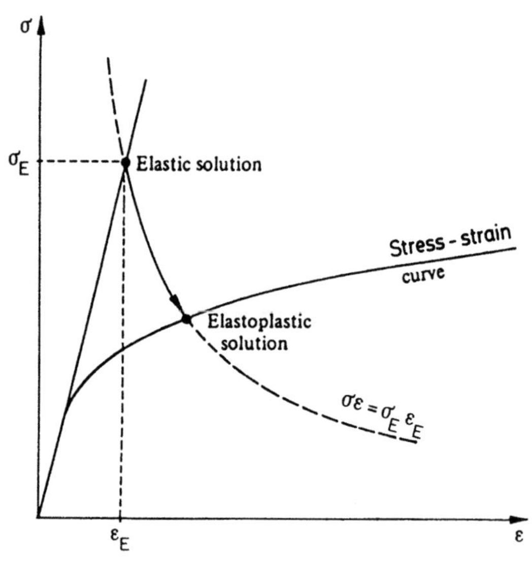

Fig. 4.12. Neuber's method

since ε_{ij}^p is a deviator: $\delta_{ij}\varepsilon_{ij}^p = 0$.

For monotonic proportional loading the plastic strain may be written as (see Section 2.1.4)

$$\varepsilon_{ij}^p = \frac{3}{2} g(\sigma_{eq}) \frac{\sigma_{ij}^D}{\sigma_{eq}},$$

where g is the hardening function, then

$$\sigma_{ij}\varepsilon_{ij}^p = \sigma_{ij}^D \frac{3}{2} g(\sigma_{eq}) \frac{\sigma_{ij}^D}{\sigma_{eq}} = g(\sigma_{eq})\sigma_{eq}.$$

The Neuber hypothesis written in three dimensions is:

$$\frac{2(1+\nu)}{3E}\sigma_{eq}^2 + \frac{3(1-2\nu)}{E}\sigma_H^2 + g(\sigma_{eq})\sigma_{eq} = \sigma_{ij}^E\varepsilon_{ij}^E.$$

But $\sigma_{ij}^E\varepsilon_{ij}^E$ may also be written as

$$\sigma_{ij}^E\varepsilon_{ij}^E = \frac{2(1+\nu)}{3E}(\sigma_{eq}^E)^2 + \frac{3(1-2\nu)}{E}(\sigma_H^E)^2;$$

then

$$\frac{2(1+\nu)}{3E}\sigma_{eq}^2 + g(\sigma_{eq})\sigma_{eq} = \frac{2(1+\nu)}{3E}(\sigma_{eq}^E)^2 + \frac{3(1-2\nu)}{E}\left[(\sigma_H^E)^2 - \sigma_H^2\right];$$

σ_{eq}^E being known from the elastic calculation, σ_{eq} may be determined if σ_H^2 (which cannot be calculated without the complete analysis) is considered close to $(\sigma_H^E)^2$ to make the last term negligible.

4.1.3 Finite Element Method (A. Benallal, R. Billardon, I. Doghri, 1988)

This method allows us to numerically solve the set of equations from problems in mechanics as described in Section 4.1. It consists in replacing the problem of partial differential equations by a problem of linear algebraic equations in which the unknowns are the displacements of the nodes defining the finite elements into which the structure is devided. In each element the continuum mechanical equations are solved for a class of simple displacement fields: linear, quadratic or of higher order with space coordinates. By consequence of the piecewise approximation of the displacement, and of the strain and stress fields, we are obliged to considerer a mesh size small enough to discretize the gradients of stress or strain.

The algebric system to be solved is linear for elasticty, the only difficulty being the size of the banded matrix for 1000, 10000, 100000 or even more degrees of freedom. When plasticity and damage occur, the problem becomes nonlinear and the only way to solve it is by timelike or load incremental linearization.

It is not the purpose of this course to describe the finite element method in detail; many good books already do. But as we are concerned with damage, two dificulties arise: the choice of the meshes and the choice of the time or load increments. Convergence regarding these two parameters is difficult to ensure and only heuristic procedures may be advised:

— The algorithms used to integrate the highly nonlinear constitutive equations must be "strong" relative to stability.
— Implicit integration schemes associated with Newton's iterative method are preferred
— The local integration of the constitutive equations is performed by using an "elastic predictor" to calculate the first increment as elastic and a "plastic corrector" to ensure, by iterations, that the plasticity criterion and kinetic laws are satisfied up to the required accuracy.
— In order to compute the tangent stiffness matrix of the structure at each iteration, a tangent modulus **J** consistent with the discretization, calculated as $\delta\sigma_{ij} = J_{ijkl}\delta\varepsilon_{kl}$ is used instead of $\dot{\sigma}_{ij} = L_{ijkl}\dot{\varepsilon}_{kl}$. This ensures a much better convergence.
— As to the space discretization is concern, the convergence regarding the mesh size is of the same order of difficulty as for classical elastoplastic calculations up to crack initiation, that is just before the localization of damage. Afterwards the problem is no longer elliptic; its solution involves strain rate discontinuities which cannot be modeled by classical finite elements.
— When a periodic loading is considered, as for fatigue, it is not possible to perform the tremendous number of increments needed for all of the thousands, or millions of cycles. A method for saving computer time involves the use of a "jump-in-cycles" procedure which avoids calculating the stress strain cycles when the process is a quasi-steady state process.

  Assume a piecewise periodic loading and consider a number of cycles Ni for which the damage is known to be Di. The structural calculation of this cycle (i) is performed several times until the stabilization is reached at each point. The coupled constitutive equations are such that for constant damage, a periodic input implies an asymptotic periodic output. Only two or three calculations are sufficient.

  This gives $\left(\frac{\delta D}{\delta N}\right)_i$ at each Gauss point of the elements of the structure, this rate will be assumed to be constant during a certain number of cycles ΔNi, which may be very large.

  The criterion for choosing ΔNi, is a certain amount of damage $\overline{\Delta D}$ at the most damaged point M^*. $\overline{\Delta D}$ must not be so large as to violate the coupling, but large enough to obtain a reasonable computer time. For applications $\overline{\Delta D} = D_{1c}/50$

is a good compromise. Then

$$\Delta Ni = \frac{\overline{\Delta D}}{\left(\frac{\delta D}{\delta N}\right)_{i(M^*)}},$$

where

$$\left(\frac{\delta D}{\delta N}\right)_{i(M^*)} = \underset{(M)}{\text{Max}} \left(\frac{\delta D}{\delta N}\right)_{i(M)}.$$

The damage at each point for the next step of the number of cycles is

$$N_{i+1} = N_i + \Delta N_i \rightarrow D_{i+1(M)} = D_{i(M)} + \left(\frac{\delta D}{\delta N}\right)_{i(M)} \Delta N_i,$$

and the process may be continued in damage increments.

When needed for variable amplitude loadings, this procedure also applies for the plastic strain, even without any damage, where the game is played with the accumulated plastic strain p. After stabilization is reached with a cyclic plastic increment $\delta p/\delta N$, a jump of ΔN cycles is performed like:

$$\Delta N = \frac{\overline{\Delta p}}{\frac{\delta p}{\delta N}},$$

where $\overline{\Delta p}$ is the best compromise between accuracy and rapidity. ε_R being the order of magnitude of the strain to rupture in pure tension, $\overline{\Delta p} = \frac{\varepsilon_R}{50}$ gives good results. Nevertheless, be careful when using those heuristic procedures!

4.1.4 Exercise on the Stress Concentration Near a Hole

Consider a plate with a central hole of small radius in comparison to the dimensions of the plate, which is loaded in tension by a "far" stress $\sigma_{22} = \sigma_2^0$ (Figure 4.13). Assuming the material to be linear elastic, *find the stress concentration factor at the boundary of the hole* $(x_1 = \pm r, x_2 = 0)$ *from Section 4.1.1*;

$$K_T = \frac{\sigma_{22}}{\sigma_2^0} = 3.$$

Considering an increase of the external load σ_2^0, plasticity may occur at the point of stress concentration. Assuming the power strain function,

$$\varepsilon = g_\varepsilon(\sigma) = \left(\frac{\sigma}{K_\varepsilon}\right)^M \quad \text{for } \sigma_{22} \geq \sigma_y,$$

where σ_y is the yield stress K_ε and M are other parameters, *determine the stress concentration factor* K_T *for* $\sigma_{22} \geq \sigma_y$ *by Neuber's method*;

$$\sigma_{22}\varepsilon_{22} = \sigma^E \varepsilon^E,$$

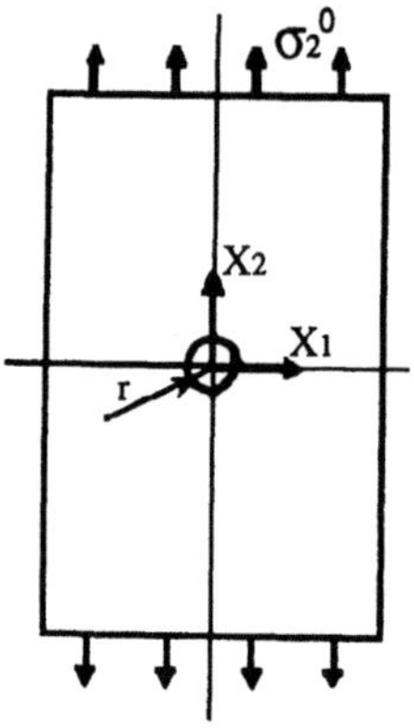

$$[\sigma]_{\binom{X_1=\pm r}{X_2=0}} = \begin{bmatrix} 0 & 0 & 0 \\ 0 & \sigma_{22} & 0 \\ 0 & 0 & 0 \end{bmatrix}$$

Fig. 4.13. Plate with a central hole

$$\sigma_{22}\left(\frac{\sigma_{22}}{K_\varepsilon}\right)^M = \frac{(3\sigma_2^0)^2}{E},$$

$$\sigma_{22} = \left(\frac{9K_\varepsilon^M}{E}\right)^{\frac{1}{M+1}} \sigma_2^{0\frac{2}{M+1}}.$$

K_T is now a function of σ_2^0 which has to be equal to 3 for $\sigma_2^0 = \dfrac{\sigma_y}{3}$. Then

$$K_T = \frac{\sigma_{22}}{\sigma_2^0} = 3\left(\frac{\sigma_y}{3\sigma_2^0}\right)^{\frac{M-1}{M+1}}.$$

Draw the graph of K_T as a function of $\dfrac{\sigma_2^0}{\sigma_y}$ for $M = 3$ and for the limiting case of perfect plasticity $M = \infty$ (Figure 4.14).

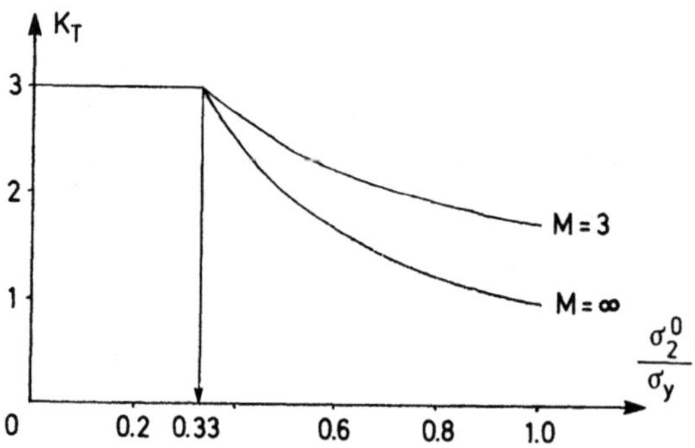

Fig. 4.14. Evolution of the stress concentration factor with plasticity

4.2 Uncoupled Analysis of Crack Initiation

Damage analysis of a structure or component subjected to a given history of loading consists in the calculation of the evolution of the damage as a function of the time at the most loaded point(s) and the critical time at which the damage reaches its critical value corresponding to mesocrack initiation.

An assumption which simplifies the analysis consists in neglecting the coupling between the damage and the strains. In a first step the stress and strain field histories are calculated by a classical method. Let us suppose that the results are

$$\varepsilon(M,t),$$
$$\sigma(M,t)$$

everywhere in the structure loaded by external forces given as a function of time.

4.2.1 Determination of the Critical Point(s)

The second step consists in determining the point(s) where the damage has its maximum value. If the loading is proportional, M^* is such a point where the damage equivalent stress is maximum at any time:

$$\sigma^*(M^*) = \underset{M}{\mathrm{Max}}(\sigma^*(M)) \rightarrow M^*,$$

$$\sigma^* = \sigma_{eq} R_\nu^{1/2}, \quad R_\nu = \frac{2}{3}(1+\nu) + 3(1-2\nu)\left(\frac{\sigma_H}{\sigma_{eq}}\right)^2$$

(see Section 2.1.2)

If the loading is nonproportional, the damage equivalent stresss may vary differently at different points as a fonction of time. A small σ^* for a long time may be more damaging than a large σ^* for a short time but less damaging than a very large σ^* as an overload! There is no rigorous way to select the critical point(s) except to calculate the evolution of the damage field in each point. This is possible, but usually, an ''intelligent'' look at the evolution of σ^* as a function of space and time will restrict the number of dangerous areas where the damage must be calculated.

4.2.2 Integration of the Kinetic Damage Law

Assume that at the critical point the structural calculation gives

$$\varepsilon_{ij}(M^*,t), \quad \varepsilon_{ij}^e(M^*,t), \quad \varepsilon_{ij}^p(M^*,t),$$
$$\sigma_{ij}(M^*,t).$$

It is easy to deduce:

— the accumulated plastic strain rate,

$$\dot{p}_{(t)} = \left(\frac{2}{3}\dot{\varepsilon}_{ij}^p \dot{\varepsilon}_{ij}^p\right)^{1/2};$$

— the strain energy density release rate,

$$Y_{(t)} = \frac{\sigma_{eq}^2 R_\nu}{2E(1-D)^2},$$

from

$$\sigma_{(eq)} = \left(\frac{3}{2}\sigma_{ij}^D \sigma_{ij}^D\right)^{1/2},$$

$$R_\nu = \frac{2}{3}(1+\nu) + 3(1-2\nu)\left(\frac{\sigma_H}{\sigma_{eq}}\right)^2,$$

$$\sigma_H = \frac{1}{3}\sigma_{kk};$$

and to write the general kinetic damage law as

$$\dot{D} = \frac{Y(t,D)}{S}\dot{p}(t)H(p-p_D).$$

or

$$\dot{D} = \frac{\sigma_{eq}^2(t)R_\nu(t)}{2ES(1-D)^2}\dot{p}(t)H(p-p_D).$$

The initial condition for the damage evolution is the end of the period of incubation of micro defects corresponding to the value p_D of the accumulated plastic strain (see Section 3.1.2). Let us start the process with the initial conditions

$$t = 0 \rightarrow p = 0,$$

$$t = t_0 \rightarrow \begin{cases} p & = p_D = \varepsilon_{PD}\dfrac{\sigma_u - \sigma_f}{\mathrm{Max}(\sigma_{eq}) - \sigma_f} \\ D & = 0 \end{cases}.$$

The damage evolution is given by the integral

$$\int_0^D (1-D)^2\, dD = \frac{1}{2ES}\int_{t_0}^t \sigma_{eq}^2(t)R_\nu(t)\dot{p}(t)\, dt$$

or

$$\frac{1}{3}\left[-(1-D)^3 + 1\right] = \frac{1}{2ES}\int_{t_0}^t \sigma_{eq}^2(t)R_\nu(t)\dot{p}(t)\, dt$$

or

$$\boxed{D = 1 - \left[1 - \frac{3}{2ES}\int_{t_0}^t \sigma_{eq}^2(t)R_\nu(t)\dot{p}(t)\, dt\right]^{1/3}}.$$

The critical time t_R at which a crack is initiated is reached when the damage itself reaches its critical value given by the rupture criterion (see Section 3.1.3)

$$D_c = D_{1c}\frac{\sigma_u^2}{\tilde{\sigma}^{*2}} \leq 1.$$

Let us take here for simplicity

$$D_c = 1.$$

t_R is given by the expression

$$1 = 1 - \left[1 - \frac{3}{2ES} \int_{t_0}^{t_R} \sigma_{eq(t)}^2 R_{\nu(t)} \dot{p}_{(t)} \, dt \right]$$

or

$$\boxed{\int_{t_0}^{t_R} \sigma_{eq(t)}^2 R_{\nu(t)} \dot{p}_{(t)} \, dt = \frac{2}{3} ES,}$$

where t_0 is given by the elastoplastic analysis: $t_0 = t(p = p_D)$

Lower bound on the life time

In the case of strain-controlled conditions it is possible to demonstrate that the last expression is in fact a lower bound on the critical time to rupture in comparison with the strain-damage coupled calculation.

Assume at first the material to be elastic-perfectly plastic and the loading to be proportional:

$\sigma_{eq} = \sigma_s = \text{const}\,.\forall t,$

$R_\nu = \text{const}\,.\forall t.$

The last expression becomes

$$\int_{t_0}^{t_R} \dot{p}(t) \, dt = \frac{2}{3} \frac{ES}{\sigma_s^2 R_\nu}.$$

The strain history is imposed on the mesovolume element considered; this allows us to take $\dot{p}(t)$ as a given function, which is particularly simple to calculate if the elastic strain is neglected:

$$\dot{p}(t) = \left(\frac{2}{3} \dot{\varepsilon}_{ij(t)} \dot{\varepsilon}_{ij(t)} \right)^{1/2}.$$

Then t_R may be determined.

Let us now consider the coupled case of the same material loaded under the same conditions:

— for $p < p_D$ no damage occurs; the same calculation as previously gives the same result,

$$t_0 = t(p = p_D);$$

— the material is perfectly plastic with the same threshold σ_s, which allows us to write the coupled plasticity criterion as:

$$\frac{\sigma_{eq}}{1 - D} = \sigma_s = \text{const}\,.;$$

— the loading is proportional:

$$R_\nu = \text{const}\,.;$$

— the same strain history is imposed and the elastic strain is again neglected: the function $\dot{p}(t)$ is the same.

The critical time for crack initiation in the coupled case t'_R is deduced from the same kinetic damage law:

$$\dot{D} = \frac{\sigma_{eq}^2 R_\nu}{2ES(1-D)^2}\dot{p}(t)H_{(p-p_D)}$$

or

$$\dot{D} = \frac{\sigma_s^2 R_\nu}{2ES}\dot{p}(t)H_{(p-p_D)},$$

where $(1-D)^2$ disappears due to the coupling in the plasticity criterion. The integration is obvious:

$$D = \frac{\sigma_s^2 R_\nu}{2ES}\int_{t_0}^{t} \dot{p}(t)\,dt.$$

Taking again $D = 1$ as the critical condition:
$D = 1 \rightarrow t = t'_R$,

$$\int_{t_0}^{t'_R} \dot{p}(t)\,dt = \frac{2ES}{\sigma_s^2 R_\nu}.$$

Comparison with the uncoupled case shows that

$$\int_{t_0}^{t'_R} \dot{p}(t)\,dt = 3\int_{t_0}^{t_R} \dot{p}(t)\,dt,$$

which implies that
$t_R < t'_R$.
In the particular case of monotonic loading where $\dot{p} = \text{const.}$,

Uncoupled case	Coupled case
$t_R = t_0 + \frac{2}{3}\frac{ES}{\sigma_s^2 R_\nu \dot{p}}$	$t'_R = t_0 + 2\frac{ES}{\sigma_s^2 R_\nu \dot{p}}$

The uncoupled calculation is always conservative. It is a lower bound, but far from the coupled solution. In other terms, component design may benefit from coupled calculations, which can prove the enhanced safety of components or indicate how light-weight economical components could be built.
The scheme of the uncoupled method is sketched in Figure 4.15.

4.2.3 Exercise on Fatigue Crack Initiation Near a Hole

Consider the same plate with a central hole as in Exercise 4.14, but now loaded with a stress at infinity varying with the number of cycles to give rise to a low cycle fatigue process:

$$\sigma^0_{2(t)} = \frac{S}{2}N\sin\omega t,$$

where N is the number of cycles (Figure 4.16).
Determine the number of cycles to crack initiation.
The whole process may be divided into three steps:

1. No plasticity, no damage until the number of cycles reaches the value N_y for which the maximum stress concentration $\sigma_M = \Delta\sigma/2$ equals the yield stress σ_y.
2. Plasticity occurs but no damage until the number of cycles reaches the value N_0 for which the accumulated plastic strain is equal to the damage threshold p_D.
3. Plasticity and damage occur to develop a damage process up to the critical value corresponding to the number of cycles to crack initiation N_R.

Step 1: $0 \leq N < N_y$.
The material is elastic everywhere; from exercise 4.1.4, the stress concentration factor is

$$K_T = 3.$$

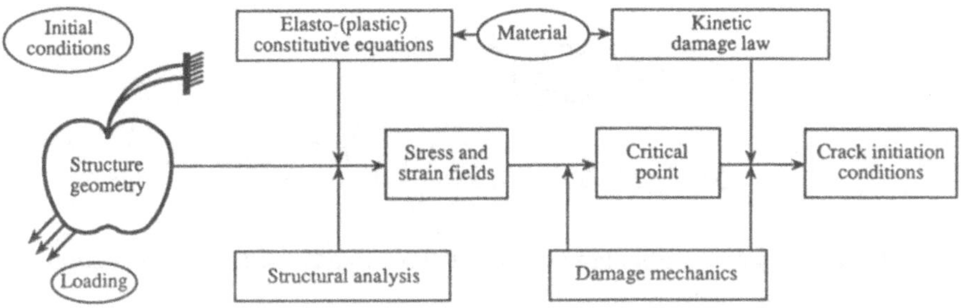

Fig. 4.15. Uncoupled analysis of crack initiation

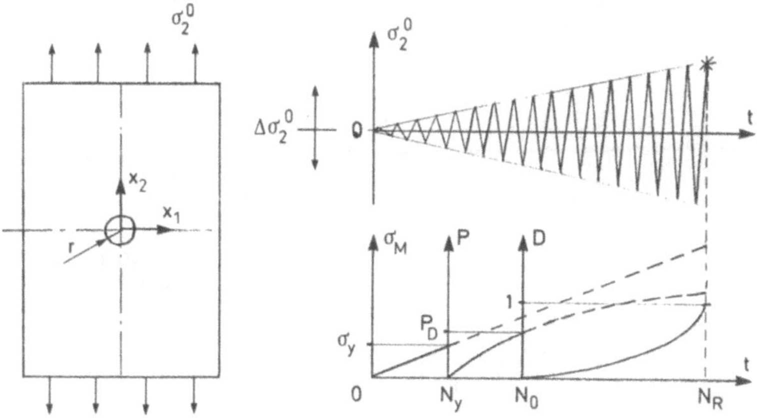

Fig. 4.16. Low cycle fatigue of a plate with a central hole

The loading history is such that

$$\Delta\sigma_2^0 = sN.$$

From the Neuber method, Section 4.1.2,

$$\Delta\sigma = K_T \Delta\sigma_2^0 = 3sN.$$

The number of cycles N_y corresponds to $\Delta\sigma = 2\sigma_y$:

$$N_y = \frac{2\sigma_y}{3s}.$$

Step 2: $N_y \leq N < N_0$.
The material is elastoplastic at the free boundary of the hole ($x_1 = \pm r$, $x_2 = 0$). Then the stress concentration factor is now a function of the stress at infinity and by consequence, a function of the number of cycles. For a cyclic loading (Section 4.1.2 and Exercise 4.1.4),

$$K_T = 3\left(\frac{2\sigma_y}{3\Delta\sigma_2^0}\right)^{\frac{M-1}{M+1}},$$

$$\Delta\sigma = K_T \Delta\sigma_2^0 = 3s\left(\frac{2\sigma_y}{3s}\right)^{\frac{M-1}{M+1}} N^{\frac{2}{M+1}}.$$

Using the cyclic strain-hardening power function already mentioned in Section 3.4.1 (here without any damage),

$$\Delta\varepsilon_p = \left(\frac{\Delta\sigma}{K_p}\right)^M.$$

The damage threshold p_D is determined by the accumulation of $2\Delta\varepsilon_p$ at each cycle. Take $p_D = \varepsilon_{pD}$ for simplicity:

$$2\int_{N_y}^{N_0} \Delta\varepsilon_{p(N)}\, dN = p_D = \varepsilon_{pD}$$

from which

$$N_0 = \left(N_y^{\frac{3M+1}{M+1}} + \varepsilon_{pD}\frac{3M+1}{2(M+1)}\left[\frac{3s\left(\frac{2\sigma_y}{3s}\right)^{\frac{M-1}{M+1}}}{K_p}\right]^{-M}\right)^{\frac{M+1}{3M+1}}.$$

If you work with numbers for a specific material this formula will not look so bad!

Step 3: $N_0 \leq N \leq N_R$.
A low cycle fatigue damage occurs with a varying amplitude of stress. Take the damage per cycle derived in Section 3.4.1:

$$\frac{\delta D}{\delta N} = \frac{K_p^2}{4ES} \Delta \varepsilon_p^{\frac{M+2}{M}} .$$

The uncoupled hypothesis assumes the plastic strain amplitude to be related to the stress amplitude without any consideration of damage:

$$\Delta \varepsilon_p = \left(\frac{\Delta \sigma}{K_p} \right)^M ,$$

or, with the relation of $\Delta \sigma$ derived in step 2,

$$\Delta \varepsilon_p = \left[\frac{3s \left(\frac{2\sigma_y}{3s} \right)^{\frac{M-1}{M+1}}}{K_p} \right]^M N^{\frac{2M}{M+1}} .$$

The number of cycles to crack initiation is obtained by integration of the kinetic law of damage:

$$\frac{\delta D}{\delta N} = \frac{K_p^2}{4ES} \left[\frac{3s \left(\frac{2\sigma_y}{3s} \right)^{\frac{M-1}{M+1}}}{K_p} \right]^{M+2} N^{\frac{2(M+2)}{M+1}} ,$$

from the initial condition $N = N_0 \rightarrow D = 0$;
to the final condition $D = D_c \rightarrow N = N_R$.
Taking again for simplicity $D_c = 1$,

$$\int_0^1 \delta D = \frac{K_p^2}{4ES} \left[\frac{3s \left(\frac{2\sigma_y}{3s} \right)^{\frac{M-1}{M+1}}}{K_p} \right]^{M+2} \int_{N_0}^{N_R} N^{\frac{2(M+2)}{M+1}} \delta N ,$$

from which it follows that

$$N_R = \left(N_0^{\frac{3M+5}{M+1}} + \frac{4ES}{K_p^2} \left[\frac{3s \left(\frac{2\sigma_y}{3s} \right)^{\frac{M-1}{M+1}}}{K_p} \right]^{-(M+2)} \cdot \frac{3M+5}{M+1} \right)^{\frac{M+1}{3M+5}} .$$

4.3 Locally Coupled Analysis (J. Lemaitre, 1988)

Quite often, the damage is so localized that the volume of the damaged material is small in comparison to the macroscale of the structural component and even to the RVE. This allows us to perform an uncoupled analysis at the macroscale and to consider the coupling between strain and damage only on the RVE of the critical point as shown schematically in Figure 4.17. This is the case of small-scale damage.

The uncoupled analysis differs from the coupled calculation at the level of the constitutive equations of the critical point. The latter gives a more accurate result, but remember that the uncoupled analysis is a lower bound on the lifetime.

The method of locally coupled analysis may often be used with good accuracy for brittle and fatigue types of damage.

4.3.1 Localization of Damage

Damage localization results from stress concentration, of course, but also occurs, because some weakness always exists at the microscale. Let us generalize, for any kind of damage, what has been said for quasi-brittle materials and for high cycle fatigue in Section 3.2.2 and 3.4.3. The mechanical model was, a two-scale volume element, elastic or elastoplastic at the mesoscale and elastoplastic and damageable at the microscale (Figure 3.8).

The only material characteristics which differ in the matrix and in the inclusion is the yield stress of the inclusion σ_y^μ, which is assumed to be equal to the fatique limit σ_f of the material, and its fatigue limit, assumed to be reduced in the same proportion:

$$\sigma_y^\mu = \sigma_f, \quad \sigma_f^\mu = \sigma_f \frac{\sigma_f}{\sigma_y} = \frac{\sigma_f^2}{\sigma_y}.$$

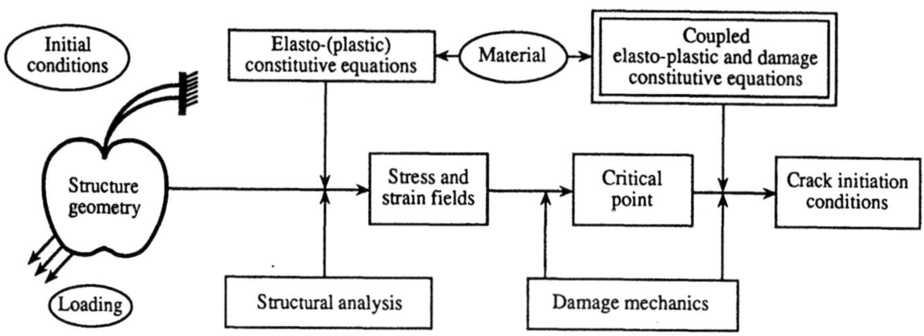

Fig. 4.17. Locally coupled analysis of crack initiation

This takes into account the microinternal stresses and the weak defects always existing everywhere in all materials. As the fatigue limit is smaller than the yield stress, it allows for plasticity which induces damage at the microscale, whereas the matrix remains elastic or elastoplastic and undamaged for the same loading.

The second assumption which simplifies calculations is the Lin-Taylor strain compatibility hypothesis, which states that the state of strain at the microscale is equal to the state of strain at the mesoscale as derived from the classical structural calculation,

$$\boldsymbol{\varepsilon}^{\mu} = \boldsymbol{\varepsilon}$$

Then, there is no boundary value problem to be considered. Only the set of coupled constitutive equations must be solved for the given history of the strain.

As in Section 3.2.2, the crack initiation condition at the microscale $D = D_C$ is also the brittle crack initiation at the mesoscale $G = G_C$.

4.3.2 Postprocessing of Damage Growth (C. Lienard, 1989; I. Doghri, 1990)

As the input of the locally coupled method is the strain at some point, it can be set up as a postprocessor to any finite element code of structural calculations. The most sophisticated constitutive equations to be used are:

— the set of equations developed in Section 2.1.4 for linear elasticity and for isotropic and kinematic plastic hardening,

— associated with the kinetic damage law developed in Section 3.1;

— in the case of unilateral damage microcrack closure conditions as explained in Section 2.3.4.

Such a postprocessor exists in the French Framasoft + CSI ≪SYSTUS≫ Finite Element Code.

It is also possible to develop a simpler code by assuming the material to be perfectly plastic at the micro-scale and by neglecting the microcrack clôsure effect. The equations are those of Section 3.1.4 with the conditions of quasi-brittle damage developed in Section 3.2.2.

Let us recall that

$$\varepsilon_{ij} = \varepsilon_{ij}^{e} + \varepsilon_{ij}^{p},$$

$$\varepsilon_{ij}^{e} = \frac{1+\nu}{E}\frac{\sigma_{ij}}{1-D} - \frac{\nu}{E}\frac{\sigma_{kk}}{1-D}\delta_{ij},$$

$$\dot{\varepsilon}_{ij}^{p} = \frac{3}{2}\frac{\tilde{\sigma}_{ij}^{D}}{\sigma_s^{\mu}}\dot{p} \quad \text{if } f = \frac{\sigma_{eq}}{1-D} - \sigma_s^{\mu} = 0 \text{ and } \dot{f} = 0; \text{ if not, } \dot{\varepsilon}_{ij}^{p} = 0,$$

$$\dot{D} = \frac{\sigma_s^{\mu 2}}{2ES}R_{\nu}\dot{p} \quad \text{if } p \geq p_D = \varepsilon_{pD}\frac{\sigma_u - \sigma_f}{\sigma_s^{\mu} - \dfrac{\sigma_f^2}{\sigma_y}}; \text{ if not, } \dot{D} = 0;$$

Crack initiation if $D_c = D_{1c}\dfrac{\sigma_u^2}{\sigma_s^{\mu 2} R_\nu}$.

These equations may be used for piecewise perfect plasticity by considering several values of the plastic threshold σ_s^μ as the loading or the timelike parameter vary. This allows us to take some cases of high values of strainhardening and the cyclic stress strain curve for multilevel fatigue processes into consideration.

Then, the material parameters must be considered as follows:

* E and ν for elasticity;
* σ_f, σ_y and σ_u for plasticity:
* $\sigma_s^\mu = \sigma_f$ for pure plasticity;
* $\sigma_s^\mu = \sigma_{si}$ given as input for piecewise plasticity;
* S, ε_{PD}, D_{1c} for damage.

The input of the calculation is the time history of the strain components $\varepsilon_{ij(t)}$ which may come from the result of a finite element structural calculation or be given directly. Note that the set of equations does not depend explicity upon time. (This would not be the case for viscoplasticity). The reference for history is a timelike parameter which accounts for the succession of increments of the loading only.

The outputs are:

— the timelike parameter t,

— the damage evolution $D(t)$, the last point corresponding to $D = D_c$, that is, crack initiation,

— the accumulated plastic strain evolution $p(t)$;

— the evolution of the von Mises equivalent stress $\sigma_{eq}(t)$ and the damage equivalent stress $\sigma^*(t)$;

— the evolution of the stress components $\sigma_{ij}(t)$ and the input strains $\varepsilon_{ij}(t)$.

The numerical procedure is a strain-driven incremental timelike algorithm using an elastic predictor and a plastic corrector. The hypothesis of perfect plasticity, even coupled with the softening effect of damage, allows us to explicitly formulate this plastic correction. The constitutive equations are discretized in an incremental form corresponding to a fully implicit integration scheme having unconditional stability; but since the unknowns are explicitly updated, it has the advantage of an explicit scheme in which no linear system needs to be solved.

After an elastic increment given by the law of elasticity, the plasticity criterion $f \leq 0$ is checked. If $f > 0$, the plasticity corrections are obtained by Newton's iterative procedure applied to a system of two equations deduced from the constitutive equations:

$$f = \tilde{\sigma}_{eq} - \sigma_s^\mu = 0,$$

$$h_{ij} = \tilde{\sigma}_{ij} - \lambda\varepsilon_{kk}\delta_{ij} - 2\mu(\varepsilon_{ij} - \varepsilon^{p}_{ijn}) + 2\mu\frac{3}{2}\frac{\tilde{\sigma}^{D}_{ij}}{\tilde{\sigma}_{eq}}\Delta p,$$

with

$$\lambda = \frac{E\nu}{(1-2\nu)(1+\nu)} \qquad \mu = \frac{E}{2(1+\nu)}.$$

ε^{p}_{ijn} is the plastic strain at the beginning of the increment, Δp is accumulated plastic strain increment, and the other quantities are the values at the end of the increment.

The iteration operates on

$$f + \frac{\partial f}{\partial \tilde{\sigma}_{ij}} C^{\sigma}_{ij} = 0,$$

$$h_{ij} + \frac{\partial h_{ij}}{\partial \tilde{\sigma}_{kl}} C^{\sigma}_{kl} + \frac{\partial h_{ij}}{\partial p} C_p = 0,$$

where C^{σ}_{ij} and C_p may be explicitly calculated. The resulting expressions are

$$C_p = \frac{f - N_{ij}h_{ij}}{3\mu}, \quad \text{with } N_{ij} = \frac{3}{2}\frac{\tilde{\sigma}^{D}_{ij}}{\tilde{\sigma}_{eq}},$$

$$C^{\sigma}_{ij} = -\frac{2}{3}(f - N_{kl}h_{kl})N_{ij} - \frac{1}{1 + \dfrac{3\mu}{\tilde{\sigma}_{eq}}\Delta p}\left[h_{ij} + \frac{2\mu}{\tilde{\sigma}_{eq}}\Delta p(N_{kl}h_{kl})N_{ij}\right].$$

4.3.3 Desciption and Listing of the Postprocessor DAMAGE 90

(I. Doghri, 1990)

DAMAGE 90 is a friendly computer code with fewer than 600 FORTRAN instructions; it calculates the evolution of damage up to crack initiation with the method described in Section 4.3.2. It may be used in an interactive way either as a postprocessor of the strain output of a finite element code or as an autonomous program. It is written in FORTRAN 77 as available on a Digital VAX computer. The CPU time for one complete execution is a few seconds. It distinguishes between two loading cases:

— general loading history where the history of loading is defined by the values of the strain components at given timelike parameter values. DAMAGE 90 interpolates linearly between these values;

— piecewise periodic loading for which the loading is a certain number of blocks of cycles defined by the two consecutive maximum and minimum set of strain components and the number of cycles in each block. The strain interpolations are also linear. For a large number of cycles, the jump-in-cycles procedure (which allows "jumps" of cycles to save computer time as explaned in Section 4.1.3) may be used. It is capable of accounting for initial values of damage D_0 and plastic strain p_0.

The questions asked by DAMAGE 90 to introduce the input data and the charts of the output data as they appear on the screen of the terminal are given below. (The subscript μ is omitted $\sigma_s^\mu \equiv \sigma_s$).

The complete listing is also given hereafter for those of you who wish to play with DAMAGE 90.

DAMAGE 90 Input Data

— Example of the first exercise in Section 4.3.4.

```
 ****************************************************
    Give material constants and the strains history

                 *** DAMAGE90 ***

 will give you the damage growth up to crack initiation
 ****************************************************

** ELASTICITY . Give   :
     YOUNG's modulus :
200.E+3
     POISSON's ratio :
0.3
** PERFECT PLASTICITY : plastic threshold SIGs given with loading . Give :
     Fatigue limit SIGf    :
200.
     Yield stress SIGy     :
300.
     Ultimate stress SIGu :
500.
** DAMAGE EVOLUTION :    dD = (Y/S) dp . Give : S
0.06
** DAMAGE THRESHOLD : dD=0 if p<pD. Do you know pD?   'Y' or 'N'
Y
     Give the value of pD :
10.E-2
** CRACK INITIATION : D=Dc . Do you know Dc ? 'Y'  or 'N'
Y
     Give the value of Dc (remember : 0 < Dc <1) :
0.99
** INITIAL CONDITIONS . Give :
     The value of po   :
0.
     The value of Do   :
0.
** LOADING
     Is the stress state uniaxial ?  'Y' or 'N'
Y
     Is the strain history cyclic? 'Y' or 'N'
N
```

```
***** YOUR LOADING IS NOT CYCLIC.

     Give the number of points which define the history  (200 max) :
5
     Give the values of time at these points :
0. , 1. , 2. , 3. , 4.
   Give the values of 11 -strain at these times :
0. , 0.25E-2 , 1.5E-2 , 5.E-2 ,25.E-2
     Give the values of the plastic threshold stress SIGs at these times :
200. , 300. , 400. , 500. , 500.
     Suggest an initial time increment  <  to  intervalle between first 2 points
0.1
*** CRACK INITIATION.
*** THE JOB IS ENDED. YOUR RESULTS FILES ARE :

direct.out            shear.out
FORTRAN STOP
CLR>>
```

— Example of a low cycle fatigue tension case at the macroscale strain-imposed $\varepsilon_{11} = \pm 0.425\%$, with material data of the aluminum alloy considered in the exercises in Section 4.3.4.

```
  ******************************************************
     Give material constants and the strains history

                 *** DAMAGE90 ***

  will give you the damage growth up to crack initiation
  ******************************************************

 ** ELASTICITY . Give  :
      YOUNG's modulus :
 72.E+3
      POISSON's ratio :
 0.32
 ** PERFECT PLASTICITY : plastic threshold SIGs given with loading . Give :
      Fatigue limit SIGf    :
 303.
      Yield stress SIGy     :
 306.
      Ultimate stress SIGu :
 500.
 ** DAMAGE EVOLUTION :    dD = (Y/S) dp . Give : S
 6.
 ** DAMAGE THRESHOLD : dD=0 if p<pD. Do you know pD?   'Y' or 'N'
 N
     Give EpD in tension :   pD=EpD*(SIGu-SIGf)/[SIGs-(SIGf**2/SIGy)]
 10.E-2
 ** CRACK INITIATION : D=Dc . Do you know Dc ? 'Y'  or 'N'
 N
     Give D1c in tension :  Dc = D1c * [(SIGu/SIGs)**2] / Rv
 0.99
 ** INITIAL CONDITIONS . Give :
      The value of po  :
 0.
      The value of Do  :
 0.
 ** LOADING
      Is the stress state uniaxial ?  'Y' or 'N'
 N
      Is the strain history cyclic? 'Y' or 'N'
 Y
```

```
***** YOUR LOADING IS CYCLIC.

   Do you wish the jump in cycles procedure for  large N ? 'Y' or 'N'
Y
   Give the number of blocks of constant amplitude  (50 max)
1
   Give the number of cycles for block :           1
120000
  Give for each strain component MAX and min values :
 1st peak   ,  2nd peak    of 11-strain
0.425E-2 , -0.425E-2
  Give for each strain component MAX and min values :
 1st peak   ,  2nd peak    of 22-strain
-0.136E-2 , 0.136E-2
  Give for each strain component MAX and min values :
 1st peak   ,  2nd peak    of 33-strain
-0.136E-2 , 0.136E-2
  Give for each strain component MAX and min values :
 1st peak   ,  2nd peak    of 12-strain
 0.        , 0.
  Give for each strain component MAX and min values :
 1st peak   ,  2nd peak    of 13-strain
 0.        , 0.
  Give for each strain component MAX and min values :
 1st peak   ,  2nd peak    of 23-strain
 0.        , 0.
       Give the value of the plastic threshold stress  SIGs for this block :
303.
       Suggest a number of increments per cycle  (minimum : 4) :
4

*** CRACK INITIATION.
*** THE JOB IS ENDED. YOUR RESULTS FILES ARE :

direct.out           shear.out           fatigue.out
FORTRAN STOP
CLR>>
```

DAMAGE 90 Output Data

— Partial results of example of the first exercise in Section 4.3.4.

TIME	DAMAGE	p	MISES EQ.STRESS	DAMAGE EQ.STRESS
0.0000E+00	0.0000E+00	0.0000E+00	0.0000E+00	0.0000E+00
0.1000E+00	0.0000E+00	0.0000E+00	0.5000E+02	0.5000E+02
0.2100E+00	0.0000E+00	0.0000E+00	0.1050E+03	0.1050E+03
0.3310E+00	0.0000E+00	0.0000E+00	0.1655E+03	0.1655E+03
0.4641E+00	0.0000E+00	0.1603E-03	0.2000E+03	0.2000E+03
0.6105E+00	0.0000E+00	0.5263E-03	0.2000E+03	0.2000E+03
0.7716E+00	0.0000E+00	0.9289E-03	0.2000E+03	0.2000E+03
0.9487E+00	0.0000E+00	0.1372E-02	0.2000E+03	0.2000E+03
0.1000E+01	0.0000E+00	0.1500E-02	0.2000E+03	0.2000E+03
0.1056E+01	0.0000E+00	0.1705E-02	0.3000E+03	0.3000E+03
0.1118E+01	0.0000E+00	0.2481E-02	0.3000E+03	0.3000E+03
0.1187E+01	0.0000E+00	0.3334E-02	0.3000E+03	0.3000E+03
0.1262E+01	0.0000E+00	0.4273E-02	0.3000E+03	0.3000E+03
0.1344E+01	0.0000E+00	0.5305E-02	0.3000E+03	0.3000E+03
0.1435E+01	0.0000E+00	0.6441E-02	0.3000E+03	0.3000E+03
0.1535E+01	0.0000E+00	0.7690E-02	0.3000E+03	0.3000E+03
0.1645E+01	0.0000E+00	0.9064E-02	0.3000E+03	0.3000E+03
0.1766E+01	0.0000E+00	0.1058E-01	0.3000E+03	0.3000E+03
0.1899E+01	0.0000E+00	0.1224E-01	0.3000E+03	0.3000E+03
0.2000E+01	0.0000E+00	0.1350E-01	0.3000E+03	0.3000E+03
0.2111E+01	0.0000E+00	0.1689E-01	0.4000E+03	0.4000E+03
0.2233E+01	0.0000E+00	0.2116E-01	0.4000E+03	0.4000E+03
0.2368E+01	0.0000E+00	0.2586E-01	0.4000E+03	0.4000E+03
0.2515E+01	0.0000E+00	0.3104E-01	0.4000E+03	0.4000E+03
0.2605E+01	0.0000E+00	0.3417E-01	0.4000E+03	0.4000E+03
0.2703E+01	0.0000E+00	0.3761E-01	0.4000E+03	0.4000E+03
0.2811E+01	0.0000E+00	0.4140E-01	0.4000E+03	0.4000E+03
0.2930E+01	0.0000E+00	0.4556E-01	0.4000E+03	0.4000E+03
0.3000E+01	0.0000E+00	0.4800E-01	0.4000E+03	0.4000E+03
0.3023E+01	0.0000E+00	0.5213E-01	0.5000E+03	0.5000E+03
0.3049E+01	0.0000E+00	0.5723E-01	0.5000E+03	0.5000E+03
0.3077E+01	0.0000E+00	0.6284E-01	0.5000E+03	0.5000E+03
0.3108E+01	0.0000E+00	0.6901E-01	0.5000E+03	0.5000E+03
0.3126E+01	0.0000E+00	0.7274E-01	0.5000E+03	0.5000E+03
0.3147E+01	0.0000E+00	0.7684E-01	0.5000E+03	0.5000E+03
0.3169E+01	0.0000E+00	0.8136E-01	0.5000E+03	0.5000E+03
0.3194E+01	0.0000E+00	0.8633E-01	0.5000E+03	0.5000E+03
0.3221E+01	0.0000E+00	0.9179E-01	0.5000E+03	0.5000E+03
0.3252E+01	0.0000E+00	0.9780E-01	0.5000E+03	0.5000E+03
0.3270E+01	0.3788E-01	0.1014E+00	0.4811E+03	0.4811E+03
0.3290E+01	0.7954E-01	0.1054E+00	0.4602E+03	0.4602E+03
0.3312E+01	0.1254E+00	0.1098E+00	0.4373E+03	0.4373E+03
0.3336E+01	0.1758E+00	0.1147E+00	0.4121E+03	0.4121E+03
0.3362E+01	0.2312E+00	0.1200E+00	0.3844E+03	0.3844E+03
0.3392E+01	0.2922E+00	0.1259E+00	0.3539E+03	0.3539E+03
0.3424E+01	0.3593E+00	0.1323E+00	0.3203E+03	0.3203E+03
0.3443E+01	0.3999E+00	0.1362E+00	0.3000E+03	0.3000E+03
0.3465E+01	0.4446E+00	0.1405E+00	0.2777E+03	0.2777E+03
0.3488E+01	0.4937E+00	0.1452E+00	0.2531E+03	0.2531E+03
0.3514E+01	0.5477E+00	0.1504E+00	0.2261E+03	0.2261E+03
0.3543E+01	0.6072E+00	0.1561E+00	0.1964E+03	0.1964E+03
0.3574E+01	0.6725E+00	0.1624E+00	0.1637E+03	0.1637E+03
0.3593E+01	0.7121E+00	0.1662E+00	0.1439E+03	0.1439E+03
0.3614E+01	0.7556E+00	0.1703E+00	0.1222E+03	0.1222E+03
0.3637E+01	0.8035E+00	0.1749E+00	0.9826E+02	0.9826E+02
0.3662E+01	0.8561E+00	0.1800E+00	0.7194E+02	0.7194E+02
0.3690E+01	0.9140E+00	0.1855E+00	0.4299E+02	0.4299E+02
0.3721E+01	0.9777E+00	0.1917E+00	0.1113E+02	0.1113E+02
0.3739E+01	0.1016E+01	0.1954E+00	-.8136E+01	-.8136E+01

— Partial results of the low cycle fatigue in tension $\varepsilon_{11} = \pm 0.425\%$

CYCLE	DAMAGE	p	MISES EQ.STRESS	DAMAGE EQ.STRESS
0.0000E+00	0.0000E+00	0.0000E+00	0.0000E+00	0.0000E+00
0.2500E+00	0.0000E+00	0.3667E-04	0.3030E+03	0.3034E+03
0.5000E+00	0.0000E+00	0.3667E-04	0.3000E+01	0.2814E+01
0.7500E+00	0.0000E+00	0.1100E-03	0.3030E+03	0.3034E+03
0.1000E+01	0.0000E+00	0.1100E-03	0.3000E+01	0.2814E+01
0.4547E+05	0.0000E+00	0.6669E+01	0.3030E+03	0.3034E+03
0.4547E+05	0.0000E+00	0.6669E+01	0.3000E+01	0.2814E+01
0.4547E+05	0.7811E-05	0.6670E+01	0.3030E+03	0.3034E+03
0.4547E+05	0.7811E-05	0.6670E+01	0.3000E+01	0.2814E+01
0.4547E+05	0.1562E-04	0.6670E+01	0.3030E+03	0.3034E+03
0.4548E+05	0.1562E-04	0.6670E+01	0.3000E+01	0.2814E+01
0.4640E+05	0.1450E-01	0.6806E+01	0.2986E+03	0.2990E+03
0.4640E+05	0.1450E-01	0.6806E+01	0.2957E+01	0.2773E+01
0.4640E+05	0.1450E-01	0.6806E+01	0.2986E+03	0.2990E+03
0.4640E+05	0.1450E-01	0.6806E+01	0.2957E+01	0.2773E+01
0.4640E+05	0.1451E-01	0.6806E+01	0.2986E+03	0.2990E+03
0.4640E+05	0.1451E-01	0.6806E+01	0.2956E+01	0.2773E+01
0.4640E+05	0.1451E-01	0.6806E+01	0.2986E+03	0.2990E+03
0.4640E+05	0.1451E-01	0.6806E+01	0.2956E+01	0.2773E+01
0.4733E+05	0.2899E-01	0.6942E+01	0.2942E+03	0.2946E+03
0.4733E+05	0.2899E-01	0.6942E+01	0.2913E+01	0.2733E+01
0.4733E+05	0.2899E-01	0.6942E+01	0.2942E+03	0.2946E+03
0.4733E+05	0.2899E-01	0.6942E+01	0.2913E+01	0.2733E+01
0.4733E+05	0.2900E-01	0.6942E+01	0.2942E+03	0.2946E+03
0.4733E+05	0.2900E-01	0.6942E+01	0.2913E+01	0.2733E+01
0.4733E+05	0.2901E-01	0.6942E+01	0.2942E+03	0.2946E+03
0.4733E+05	0.2901E-01	0.6942E+01	0.2913E+01	0.2733E+01
0.4826E+05	0.4349E-01	0.7078E+01	0.2898E+03	0.2902E+03
0.5569E+05	0.1595E+00	0.8168E+01	0.2547E+03	0.2550E+03
0.5569E+05	0.1595E+00	0.8168E+01	0.2522E+01	0.2365E+01
0.5662E+05	0.1740E+00	0.8304E+01	0.2503E+03	0.2506E+03
0.5662E+05	0.1740E+00	0.8304E+01	0.2478E+01	0.2325E+01
0.5662E+05	0.1740E+00	0.8305E+01	0.2503E+03	0.2506E+03
0.5662E+05	0.1740E+00	0.8305E+01	0.2478E+01	0.2325E+01
0.5662E+05	0.1740E+00	0.8305E+01	0.2503E+03	0.2506E+03
0.5662E+05	0.1740E+00	0.8305E+01	0.2478E+01	0.2325E+01
0.5662E+05	0.1740E+00	0.8305E+01	0.2503E+03	0.2506E+03
0.5662E+05	0.1740E+00	0.8305E+01	0.2478E+01	0.2325E+01
0.5755E+05	0.1885E+00	0.8441E+01	0.2459E+03	0.2462E+03
0.5755E+05	0.1885E+00	0.8441E+01	0.2435E+01	0.2284E+01
0.5755E+05	0.1885E+00	0.8441E+01	0.2459E+03	0.2462E+03
0.5755E+05	0.1885E+00	0.8441E+01	0.2435E+01	0.2284E+01
0.5755E+05	0.1885E+00	0.8441E+01	0.2459E+03	0.2462E+03
0.5755E+05	0.1885E+00	0.8441E+01	0.2435E+01	0.2284E+01
0.5755E+05	0.1885E+00	0.8441E+01	0.2459E+03	0.2462E+03
0.5755E+05	0.1885E+00	0.8441E+01	0.2435E+01	0.2284E+01
0.5848E+05	0.2030E+00	0.8577E+01	0.2415E+03	0.2418E+03
0.5848E+05	0.2030E+00	0.8577E+01	0.2391E+01	0.2243E+01
0.5848E+05	0.2030E+00	0.8577E+01	0.2415E+03	0.2418E+03
0.5848E+05	0.2030E+00	0.8577E+01	0.2391E+01	0.2243E+01
0.5848E+05	0.2030E+00	0.8577E+01	0.2415E+03	0.2418E+03
0.1077E+06	0.9713E+00	0.1580E+02	0.8686E+01	0.8696E+01
0.1077E+06	0.9713E+00	0.1580E+02	0.8600E-01	0.8067E-01
0.1086E+06	0.9858E+00	0.1593E+02	0.4298E+01	0.4303E+01
0.1086E+06	0.9858E+00	0.1593E+02	0.4255E-01	0.3992E-01
0.1086E+06	0.9858E+00	0.1593E+02	0.4298E+01	0.4303E+01
0.1086E+06	0.9858E+00	0.1593E+02	0.4255E-01	0.3992E-01
0.1086E+06	0.9858E+00	0.1593E+02	0.4296E+01	0.4301E+01
0.1086E+06	0.9858E+00	0.1593E+02	0.4253E-01	0.3990E-01
0.1086E+06	0.9858E+00	0.1593E+02	0.4293E+01	0.4298E+01
0.1086E+06	0.9858E+00	0.1593E+02	0.4251E-01	0.3987E-01
0.1096E+06	0.1000E+01	0.1607E+02	-.9480E-01	-.9492E-01

Listing of DAMAGE 90

```
C*******************************************************************
C*** DAMAGE 90
C    Elastic-perfectly plastic law coupled to a ductile damage model.
C*1* Fully implicit integration scheme.
C*2* Jump in cycles procedure.
C*** Written by  Issam Doghri
C    U.C. SANTA BARBARA, DEPT. OF MECHANICAL ENGINEERING
C*** Version : May      1990
C*******************************************************************
      IMPLICIT REAL*8 (A-H,O-Z)
      CHARACTER*1 UNIAXI,ANSPD,ANSDC,CYCLIC,JUMP,CONVER,STAB,COUPL
      CHARACTER*20 FILE1,FILE2
      CHARACTER*6 COMMENT
      PARAMETER(NTENS=6,NSTATV=11)
      REAL*8 STRS(NTENS),STATEV(NSTATV),STATEVI(NSTATV),
     . STRAN(NTENS),DSTRAN(NTENS),SIGHI(6),SIGHF(6),
     . INCUSE(50),TIM(200),HIST(6,200),EPSB(6,50),EPSS(6,50),
     . PERDIV(6),SYIEL(50)
      INTEGER ISLOPE(6),IPER(6),INDEX(6),NBCYCL(50)
      COMMON/ETIQ1/E0,XNU,SIGS,SMPD,S0
      COMMON/ETIQ2/ONE,TWO,THREE
      DATA INDEX /11,22,33,12,13,23/
C*** READ DATA
      MS=6
      MB=5
      NAXI=6
      NCYCLE=0
      SMPD=1.D+10
      DCRIT=0.99D0
      WRITE(MS,*)'  '
      WRITE(MS,*)'  '
      WRITE(MS,*)' ******************************************************
     .*'
      WRITE(MS,*)'    Give material constants and the strains history'
      WRITE(MS,*)'  '
      WRITE(MS,*)'                         *** DAMAGE90 ***'
      WRITE(MS,*)'   '
      WRITE(MS,*)' will give you the damage growth up to crack initiatio
     .n'
      WRITE(MS,*)' ******************************************************
     .*'
      WRITE(MS,*)
      WRITE(MS,*)'** ELASTICITY . Give  :'
      WRITE(MS,*)'     YOUNG''s modulus :'
      READ(MB,*)E0
      WRITE(MS,*)'     POISSON''s ratio :'
      READ(MB,*)XNU
      WRITE(MS,*)'** PERFECT PLASTICITY : plastic threshold SIGs
     . given with loading . Give :'
      WRITE(MS,*)'     Fatigue limit SIGf   :'
      READ(MB,*)SIGF
      WRITE(MS,*)'     Yield stress SIGy    :'
      READ(MB,*)SIGY
      WRITE(MS,*)'     Ultimate stress SIGu :'
      READ(MB,*)SIGU
       WRITE(MS,*)'** DAMAGE EVOLUTION :   dD = (Y/S) dp . Give : S'
       READ(MB,*)S0
 1     FORMAT(A)
       WRITE(MS,*)'** DAMAGE THRESHOLD : dD=0 if p<pD. Do you know pD?
     .   ''Y'' or ''N'''
       READ(MB,1)ANSPD
      IF((ANSPD.EQ.'y').OR.(ANSPD.EQ.'Y'))THEN
        WRITE(MS,*)'     Give the value of pD :'
        READ(MB,*)SMPD
        IF(SMPD.EQ.0.)SMPD=1.D-6
      ELSE IF((ANSPD.EQ.'n').OR.(ANSPD.EQ.'N'))THEN
```

```
     2  '      Give EpD in tension :
     3  pD=EpD*(SIGu-SIGf)/[SIGs-(SIGf**2/SIGy)]'
        READ(MB,*)EPD
        IF(EPD.EQ.0.)EPD=1.D-6
      ELSE
        STOP'*** WRONG DATA. GOOD BYE.'
      END IF
      WRITE(MS,*)'** CRACK INITIATION : D=Dc . Do you know Dc ? ''Y''
     .  or ''N'''
      READ(MB,1)ANSDC
      IF((ANSDC.EQ.'y').OR.(ANSDC.EQ.'Y'))THEN
        WRITE(MS,*)'     Give the value of Dc (remember : 0 < Dc <1) :'
        READ(MB,*)DCRIT
      ELSEIF((ANSDC.EQ.'n').OR.(ANSDC.EQ.'N'))THEN
        WRITE(MS,*)'    Give D1c in tension :
     .  Dc = D1c * [(SIGu/SIGs)**2] / Rv '
        READ(MB,*)D1C
      ELSE
        STOP'*** WRONG DATA. GOOD BYE.'
      END IF
      WRITE(MS,*)'** INITIAL CONDITIONS . Give :'
      WRITE(MS,*)'     The value of po  :'
      READ(MB,*)SMP0
      WRITE(MS,*)'     The value of Do  :'
      READ(MB,*)D0
      WRITE(MS,*)'** LOADING'
      WRITE(MS,*)'     Is the stress state uniaxial ?  ''Y'' or ''N'''
      READ(MB,1)UNIAXI
      IF(UNIAXI.NE.'y'.AND.UNIAXI.NE.'n'.AND.
     .    UNIAXI.NE.'Y'.AND.UNIAXI.NE.'N')STOP'*** WRONG DATA.
     .  GOOD BYE.***'
      IF((UNIAXI.EQ.'y').OR.(UNIAXI.EQ.'Y')) NAXI=1
      WRITE(MS,*)'     Is the strain history cyclic? ''Y'' or ''N'''
      READ(MB,1)CYCLIC
      IF(CYCLIC.NE.'y'.AND.CYCLIC.NE.'n'.AND.
     .     CYCLIC.NE.'Y'.AND.CYCLIC.NE.'N')STOP'*** WRONG DATA.
     . GOOD BYE.***'
      IF((CYCLIC.EQ.'y').OR.(CYCLIC.EQ.'Y'))THEN
      WRITE(MS,*)'  '
      WRITE(MS,*)'***** YOUR LOADING IS CYCLIC.'
      WRITE(MS,*)'  '
      WRITE(MS,*)'   Do you wish the jump in cycles procedure for
     .  large N ? ''Y'' or ''N'''
      READ(MB,1)JUMP
      IF(JUMP.NE.'y'.AND.JUMP.NE.'n'.AND.
     .    JUMP.NE.'Y'.AND.JUMP.NE.'N')STOP'*** WRONG DATA.
     .  GOOD BYE.***'
      WRITE(MS,*)'   Give the number of blocks of constant amplitude
     .  (50 max)'
      READ(MB,*)NBLOCK
      DO IBL=1,NBLOCK
      WRITE(MS,*)'   Give the number of cycles for block :',IBL
        READ(MB,*)NBCYCL(IBL)
        NCYCLE=NCYCLE+NBCYCL(IBL)
        DO IST=1,NAXI
      WRITE(MS,*)' Give for each strain component MAX and min values :'
      WRITE(MS,3)INDEX(IST)
 3    FORMAT('  1st peak   ,  2nd peak   of ',I2,'-strain')
          READ(MB,*)EPSB(IST,IBL),EPSS(IST,IBL)
        END DO
      WRITE(MS,*)'       Give the value of the plastic threshold stress
     & SIGs for this block :'
     .  READ(MB,*)SYIEL(IBL)
      WRITE(MS,*)'       Suggest a number of increments per cycle
     .  (minimum : 4) : '
        READ(MB,*)INCUSE(IBL)
      END DO
      ELSE IF((CYCLIC.EQ.'n').OR.(CYCLIC.EQ.'N'))then
      WRITE(MS,*)'  '
```

```
      WRITE(MS,*)'***** YOUR LOADING IS NOT CYCLIC.'
      WRITE(MS,*)'  '
      WRITE(MS,*)'      Give the number of points which define the histor
     .y  (200 max) :'
      READ(MB,*)NGIV
      WRITE(MS,*)'      Give the values of time at these points :'
      READ(MB,*)(TIM(IG),IG=1,NGIV)
      DO IST=1,NAXI
      WRITE(MS,2)INDEX(IST)
    2 FORMAT('    Give the values of ',I2,' -strain at these'
     . ' times :')
        READ(MB,*)(HIST(IST,IG),IG=1,NGIV)
      END DO
      WRITE(MS,*)'      Give the values of the plastic threshold stress S
     &IGs at these times :'
      READ(MB,*)(SYIEL(IG),IG=1,NGIV)
      WRITE(MS,*)'      Suggest an initial time increment  <  to
     &intervalle between first 2 points'
      READ(MB,*)DTIME
      END IF
C***INITIALIZE. STATEV= r,p,D,dD,{pstrn},dp
      DO IST=1,NSTATV
        STATEVI(IST)=0.
      END DO
      STATEVI(2)=SMP0
      STATEVI(3)=D0
      STATEVI(4)=0.
      DO IST=1,6
        STRAN(IST)=0.
        DSTRAN(IST)=0.
      END DO
      IF((ANSPD.EQ.'y').OR.(ANSPD.EQ.'Y'))EPD=SMPD
      ENERG1=(SIGU-SIGF)*EPD
      ENERG0=0.
      ENERG3=0.
      TOLENE=2.5D-2
      ILOOK0=0
      COUPL='y'
      IPASS=0
      ONE=1.D+00
      TWO=2.D+00
      THREE=3.D+00
      RAC2=DSQRT(TWO)
      NU1=11
      NU2=12
      FILE1='direct.out'
      FILE2='shear.out'
       OPEN(UNIT=NU1,name=FILE1,status='unknown')
       OPEN(UNIT=NU2,name=FILE2,status='unknown')
      COMMENT=' TIME'
      IF((CYCLIC.EQ.'y').OR.(CYCLIC.EQ.'Y'))COMMENT=' CYCLE'
      WRITE(NU1,312)COMMENT
      WRITE(NU2,312)COMMENT
      WRITE(90,315)COMMENT
  312 FORMAT(2X,A6,21X,'STRAINS',27X,'STRESSES',16X,'DAMAGE',
     . 9X,'p',8X, 'MISES',2X,'DAM.EQ.STRESS')
      WRITE(NU1,313)
      WRITE(NU2,314)
  313 FORMAT(19X,'11',9X,'22',9X,'33',10X,'11',9X,'22',9X,'33',
     .    11X,'D',23X,'SIG eq',5X,'SIG*')
  314 FORMAT(19X,'12',9X,'13',9X,'23',10X,'12',9X,'13',9X,'23',
     .    11X,'D',23X,'SIG eq',5X,'SIG*')
  315 FORMAT(5X,A6,10X,'DAMAGE',8X,'p',5X,'MISES EQ.STRESS'
     .       ,2X,'DAMAGE EQ.STRESS')
C*** IF THE LOADING IS NOT CYCLIC
      IF((CYCLIC.EQ.'n').or.(CYCLIC.EQ.'N'))THEN
        JUMP='n'
        TIME=TIM(1)
        DTMIN=DTIME/10000.
        TFIN=TIM(NGIV)
        IGIV=1
```

```
        CALL OUTPUT(NU1,NU2,TIME,STRAN,STRS,STATEV,NTENS,NSTATV,
     . STRSB,STAR)
        DOWHILE(TIME+DTIME.LE.TFIN)
          DO IST=1,NAXI
            DSTRAN(IST)=DTIME*(HIST(IST,IGIV+1)-HIST(IST,IGIV))
     &                   /(TIM(IGIV+1)-TIM(IGIV))
          END DO
          CONVER='y'
          DO IST=1,NSTATV
            STATEV(IST)=STATEVI(IST)
          END DO
          SIGS=SYIEL(IGIV)
          IF(SIGS.LT.SIGF.OR.SIGS.GT.SIGU)STOP'*** PLEASE REVIEW THE
     &    VALUES OF  SIGf  SIGu   and  SIGs .'
          CALL INTEGR(STRAN,DSTRAN,NTENS,NSTATV,UNIAXI,COUPL,
     &                CONVER,STRS,STATEV,STRSB,STAR,RNU)
C*** IF NO CONVERGENCE, DIVIDE TIME INC. BY 2
          IF(CONVER.EQ.'y'.AND.STATEV(11).GT.0.)THEN
            ENERG2=ENERG3+(SIGS-SIGF*(SIGF/SIGY))*STATEV(11)
            IF(ANSDC.EQ.'n')DCRIT=D1C*((SIGU/SIGS)**2)/RNU
            IF(ILOOK0.EQ.0.AND.(ANSPD.EQ.'y'.OR.ANSPD.EQ.'Y')
     .      .AND.STATEV(2).GT.1.05*SMPD.OR.
     .      ILOOK0.EQ.0.AND.(ANSPD.EQ.'n'.OR.ANSPD.EQ.'N').AND.
     .      ENERG2.GT.1.05*ENERG1.OR.
     .      STATEV(3).GT.1.05*DCRIT)THEN
               CONVER='n'
            END IF
          END IF
          IF(CONVER.EQ.'n')THEN
            IPASS=IPASS+1
            DTIME=DTIME/2.
C***  IF CONVERGENCE
          ELSE
            IF((ANSDC.EQ.'n'.OR.ANSDC.EQ.'N')
     .         .AND.DCRIT.GT.0.99D0)DCRIT=0.99D0
            IPASS=0
            DO IST=1,NSTATV
              STATEVI(IST)=STATEV(IST)
            END DO
            DO IST=1,6
              STRAN(IST)=STRAN(IST)+DSTRAN(IST)
            END DO
            TIME=TIME+DTIME
        CALL OUTPUT(NU1,NU2,TIME,STRAN,STRS,STATEV,NTENS,NSTATV,
     . STRSB,STAR)
            IF(STATEV(3).GE.DCRIT)THEN
               WRITE(MS,*)'  '
               WRITE(MS,*)'*** CRACK INITIATION.'
               GO TO 308
            END IF
            IF(STATEV(3).GE.1.1)STOP'*** DAMAGE CANNOT EXCEED 1.'
C*** FIND pD
             IF(ILOOK0.EQ.0.AND.STATEV(11).GT.0.)THEN
               ENERG3=ENERG2
               IF((ANSPD.EQ.'n'.OR.ANSPD.EQ.'N')
     .           .AND.ENERG3.GE.ENERG1)THEN
                 SMPD=STATEV(2)
                 ILOOK0=1
               ELSE IF((ANSPD.EQ.'y').OR.(ANSPD.EQ.'Y').AND.
     .                STATEV(2).GE.SMPD)THEN
                 ILOOK0=1
               END IF
             END IF
            IF(DABS(TIME-TIM(IGIV+1)).LE.DTMIN)THEN
              IGIV=IGIV+1
            END IF
          END IF
          DTIME=DTIME*1.1
          IF(TIME+DTIME.GT.TIM(IGIV+1)+DTMIN.AND.IGIV+1.LE.NGIV)THEN
            DTIME=TIM(IGIV+1)-TIME
          END IF
```

```
          IF(IPASS.EQ.13.OR.DTIME.LT.DTMIN)THEN
            WRITE(MS,*)'*** NO CONVERGENCE'
            WRITE(MS,*)'DTIME=',DTIME,'IPASS=',IPASS
            GO TO 308
          END IF
        END DO
      END IF
C*** IF THE LOADING IS CYCLIC
      IF((CYCLIC.EQ.'y').OR.(CYCLIC.EQ.'Y'))THEN
        TPER=1.
        DTMIN=TPER/40000.
        TIME=0.
        ICYCLE=1
        NRCYCL=ICYCLE
        DHTIME=0.
        DO I=1,6
          SIGHI(I)=0.
        END DO
        SMPI=SMP0
        DHI=D0
        IF((JUMP.EQ.'y').OR.(JUMP.EQ.'Y'))THEN
          COUPL='n'
         OPEN (UNIT=80,name='fatigue.out',status='unknown')
         WRITE(80,*)'REAL TIME          TIME          D          p'
         WRITE(80,*)'********************************'
         WRITE(80,*)'BEGINNING OF THE CYCLE'
         WRITE(80,*)'CYCLE (REAL)      =',NRCYCL
         WRITE(80,*)'CYCLE (MACHINE)   =',ICYCLE
         WRITE(80,400)RTIME,TIME,DHI,SMPI
        END IF
        CALL OUTPUT(NU1,NU2,TIME,STRAN,STRS,STATEV,NTENS,NSTATV,
     .  STRSB,STAR)
        IBL=1
        DTIME=TPER/INCUSE(IBL)
        DO IST=1,6
          IF(EPSB(IST,IBL)*EPSS(IST,IBL).LT.0.)THEN
            PERDIV(IST)=TPER/4.
          ELSE
            PERDIV(IST)=TPER/2.
          END IF
          IPER(IST)=1
          ISLOPE(IST)=1
        END DO
        NCYSUM=0
        DOWHILE(NRCYCL.LE.NCYCLE)
          DO IST=1,NAXI
           IF(TIME+DTMIN.LE.PERDIV(IST))THEN
            DSTRAN(IST)=EPSB(IST,IBL)*DTIME/PERDIV(IST)
           ELSE
           xxnu= epsb(ist,ibl)-epss(ist,ibl)
           DSTRAN(IST)=ISLOPE(IST)*xxnu*2.*DTIME/TPER
           END IF
          END DO
          CONVER='y'
          DO IST=1,NSTATV
            STATEV(IST)=STATEVI(IST)
          END DO
          SIGS=SYIEL(IBL)
          EH1=SIGS/1000.
          EPSMIN=SIGS/E0/10000.
          IF(SIGS.LT.SIGF.OR.SIGS.GT.SIGU)STOP'*** PLEASE REVIEW THE
     &    VALUES OF SIGf  SIGu   and  SIGs .'
          CALL INTEGR(STRAN,DSTRAN,NTENS,NSTATV,UNIAXI,COUPL,
     &                CONVER,STRS,STATEV,STRSB,STAR,RNU)
C*** IF NO CONVERGENCE, DIVIDE TIME INC. BY 2
          IF((ANSDC.EQ.'n'.OR.ANSDC.EQ.'N').AND.
     .        STATEV(11).GT.0.)
     .         DCRIT=D1C*((SIGU/SIGS)**2)/RNU
          IF(CONVER.EQ.'n')THEN
            IPASS=IPASS+1
            DTIME=DTIME/2.
```

```
C***  IF CONVERGENCE
          ELSE
            IF((ANSDC.EQ.'n'.OR.ANSDC.EQ.'N')
             .AND.DCRIT.GT.0.99D0)DCRIT=0.99D0
            IPASS=0
            DO IST=1,NSTATV
              STATEVI(IST)=STATEV(IST)
            END DO
            DO IST=1,6
              STRAN(IST)=STRAN(IST)+DSTRAN(IST)
            END DO
            TIME=TIME+DTIME
            RTIME=TIME+DHTIME
        CALL OUTPUT(NU1,NU2,RTIME,STRAN,STRS,STATEV,NTENS,NSTATV,
       STRSB,STAR)
            IF(STATEV(3).GE.DCRIT)THEN
              WRITE(MS,*)'  '
              WRITE(MS,*)'*** CRACK INITIATION.'
              GO TO 308
            END IF
            IF(STATEV(3).GE.1.1)STOP'*** DAMAGE CANNOT EXCEED 1.'
            DO IST=1,6
              IF(DABS(TIME-IPER(IST)*PERDIV(IST)).LE.DTMIN)THEN
                IPER(IST)=IPER(IST)+1
                DTIME=1./INCUSE(IBL)
              END IF
      IF(ISLOPE(IST).GT.0.AND.DABS(STRAN(IST)-EPSB(IST,IBL)).LE.
     . EPSMIN.OR.ISLOPE(IST).LT.0.AND.DABS(STRAN(IST)-EPSS(IST,IBL))
     . .LE.EPSMIN)THEN
                ISLOPE(IST)=-ISLOPE(IST)
      END IF
            END DO
C*** END OF A CYCLE
            IF(DABS(TIME-TPER*ICYCLE).LE.DTMIN)THEN
              IDELTN=0
              DO I=1,6
                SIGHF(I)=STRS(I)
              END DO
              SMPF=STATEV(2)
              DHF=STATEV(3)
              DELTP=SMPF-SMPI
              DELTD=DHF-DHI
C*** FIND pD
              IF(ILOOK0.EQ.0)THEN
                ENERG3=ENERG0+(SIGS-SIGF*(SIGF/SIGY))*
     .            DELTP*(NRCYCL-NCYSUM)
                IF((ANSPD.EQ.'n'.OR.ANSPD.EQ.'N')
     .            .AND.ENERG3.GE.ENERG1)THEN
                  SMPD=SMPF
                  NCYCL0=NRCYCL-NCYSUM
                  ILOOK0=1
                ELSE IF((ANSPD.EQ.'y'.OR.ANSPD.EQ.'Y').AND.
     .                STATEV(2).GE.SMPD)THEN
                  NCYCL0=NRCYCL-NCYSUM
                  ILOOK0=1
                END IF
              END IF
C*** JUMP IN CYCLES PROCEDURE
              IF((JUMP.EQ.'y').OR.(JUMP.EQ.'Y'))THEN
        WRITE(80,*)'END OF THE CYCLE'
        WRITE(80,400)RTIME,TIME,DHF,SMPF
                DDMAX0=DCRIT/50.
                IF((ANSDC.EQ.'n').OR.(ANSDC.EQ.'N'))
     .          DDMAX0=D1C/50.
                IF((UNIAXI.EQ.'y').OR.(UNIAXI.EQ.'Y'))THEN
                  YYY=SIGS*SIGS/2./E0
                ELSE
                  TRDEPS=0.
                  DO IST=1,3
                    TRDEPS=TRDEPS+EPSB(IST,IBL)-EPSS(IST,IBL)
                  END DO
```

```
               YY1=(1.+XNU)*SIGS*SIGS/3./E0
               YY2=E0*TRDEPS*TRDEPS/(1.-2.*XNU)/6.
               YYY=YY1+YY2
             END IF
             DPMAX0=S0*DDMAX0/YYY
             DPMAX=DPMAX0
             PSLOPE=NBCYCL(IBL)*DELTP
             IF(DPMAX.GT.PSLOPE)DPMAX=PSLOPE
           WRITE(80,*)'ILOOK0=',ILOOK0,'NCYCL0=',NCYCL0,'pD=',SMPD
             IF(COUPL.EQ.'n'.OR.STATEV(2).LT.SMPD)THEN
               STAB='y'
               DO KK=1,6
                 IF(DABS(SIGHF(KK)-SIGHI(KK)).GT.EH1)STAB='n'
               END DO
      WRITE(80,*)'ENERG1=',ENERG1,'ENERG3=',ENERG3
               IF(STAB.EQ.'y') THEN
      WRITE(80,*)'STABILIZED CYCLE'
                 IF(STATEV(2).LT.SMPD)THEN
C*** JUMP OF CYCLES BEFORE DAMAGE GROWTH
                   IDELTN=IDINT(DPMAX/DELTP)
      IF(NRCYCL+IDELTN.GT.NCYSUM+NBCYCL(IBL))
     .      IDELTN=NCYSUM+NBCYCL(IBL)-NRCYCL
      ENERG2=ENERG0+(SIGS-SIGY)*DELTP*(NRCYCL+IDELTN-NCYSUM)
      IF(ENERG2.GT.ENERG1*1.05)
     .      IDELTN=(ENERG1-ENERG3)/(SIGS-SIGY)/DELTP
      WRITE(80,*)'*** JUMP OF ',IDELTN,' CYCLES'
      WRITE(80,*)'DPMAX=',IDELTN*DELTP, 'YYY=',YYY
                ELSE
      WRITE(80,*)'COUPLED COMPUTATION FOR NEXT CYCLE '
                  COUPL='y'
                END IF
              END IF
            ELSE
C*** JUMP OF CYCLES AFTER DAMAGE GROWTH
              DDMAX=DDMAX0
              DSLOPE=(NBCYCL(IBL)-NCYCL0)*DELTD
              IF(DDMAX.GT.DSLOPE)DDMAX=DSLOPE
              NCYCL0=0
              IDELTN=IDINT(DDMAX/DELTD)
      IF(NRCYCL+IDELTN.GT.NCYSUM+NBCYCL(IBL))
     . IDELTN=NCYSUM+NBCYCL(IBL)-NRCYCL
      IF(DHF+DELTD*IDELTN.GT.DCRIT*1.05)
     . IDELTN=IDINT((DCRIT-DHF)/DELTD)
      IF(IDELTN.GT.IDINT(DPMAX/DELTP))IDELTN=IDINT(DPMAX/DELTP)
      WRITE(80,*)'*** JUMP OF ',IDELTN,' CYCLES'
      WRITE(80,*)'DDMAX=',IDELTN*DELTD, 'DPMAX=',IDELTN*DELTP
              COUPL='n'
            END IF
      IF(DELTP.GT.DPMAX)WRITE(80,*)'*** PROBLEM :
     .  DPMAX0 IS TOO SMALL.'
      IF(DELTD.GT.DDMAX)WRITE(80,*)'*** PROBLEM :
     . DDMAX0 IS TOO SMALL.'
      IF(DELTD.GT.DDMAX.OR.DELTP.GT.DPMAX)IDELTN=0
      IF(IDELTN.LT.0.)STOP'*** PROBLEM : NEGATIVE JUMP OF CYCLES.'
          END IF
          DO J=1,6
            SIGHI(J)=SIGHF(J)
          END DO
          SMPI=SMPF+DELTP*IDELTN
          DHI=DHF+DELTD*IDELTN
          STATEVI(2)=SMPI
          STATEVI(3)=DHI
          DHTIME=DHTIME+TPER*IDELTN
          NRCYCL=NRCYCL+IDELTN
          RTIME=TIME+DHTIME
          IF((JUMP.EQ.'y').OR.(JUMP.EQ.'Y'))THEN
      WRITE(80,*)'--------------------------'
      WRITE(80,*)'BEGINNING OF THE CYCLE'
      WRITE(80,*)'CYCLE (REAL)      =',NRCYCL+1
      WRITE(80,*)'CYCLE (MACHINE)   =',ICYCLE+1
      WRITE(80,400)RTIME,TIME,DHI,SMPI
          END IF
```

```
              IF(NRCYCL.EQ.NCYSUM+NBCYCL(IBL))THEN
           ENERG0=ENERG0+(SIGS-SIGF*(SIGF/SIGY))*DELTP*(NRCYCL-NCYSUM)
                NCYSUM=NCYSUM+NBCYCL(IBL)
                IBL=IBL+1
              END IF
              ICYCLE=ICYCLE+1
              NRCYCL=NRCYCL+1
            END IF
400       FORMAT(1X,4(E12.6,','))
          END IF
          DO IST=1,6
            IF(TIME+DTIME.GT.IPER(IST)*PERDIV(IST)+DTMIN)THEN
              DTIME=(IPER(IST)*PERDIV(IST))-TIME
            END IF
          END DO
          IF(IPASS.EQ.13.OR.DTIME.LT.DTMIN)THEN
            WRITE(MS,*)'*** NO CONVERGENCE'
            WRITE(MS,*)'DTIME=',DTIME,'IPASS=',IPASS
            GO TO 308
          END IF
        END DO
      END IF
308   WRITE(MS,*)'*** THE JOB IS ENDED. YOUR RESULTS FILES ARE :'
      WRITE(MS,*)'  '
      IF((JUMP.EQ.'n').OR.(JUMP.EQ.'N'))WRITE(MS,*)FILE1,FILE2
      IF((JUMP.EQ.'y').OR.(JUMP.EQ.'Y'))
     . WRITE(MS,*)FILE1,FILE2,'fatigue.out'
      STOP
      END
C*****************************************************************
      SUBROUTINE INTEGR(STRAN,DSTRAN,NTENS,NSTATV,UNIAXI,COUPL,
     &                CONVER,STRS,STATEV,STRSB,STAR,RNU)
C*****************************************************************
      IMPLICIT REAL*8 (A-H,O-Z)
      REAL*8 STATEV(NSTATV),STRAN(NTENS),DSTRAN(NTENS),
     &  STRN(6),ID(6,6),AS(6),
     &  STRS(6),ESTRS(6),ESTRS0(6),CESTRS(6),KESTRS(6),
     &  N(6),PSTRN(6),DPSTRN(6)
      REAL*8 LAMDA,NN
      LOGICAL CRITERE
      CHARACTER*1 CONVER,COUPL,UNIAXI
      COMMON/ETIQ1/E0,XNU,SIGS,SMPD,S0
      COMMON/ETIQ2/ONE,TWO,THREE
C
      NN=THREE/TWO
      RAC2=DSQRT(TWO)
      TOLRES=SIGS*1.D-6
      NISS=NTENS
C*** IDENTITY MATRIX
      DO I=1,NISS-1
        DO J=I+1,NISS
          ID(I,J)=0.
          ID(J,I)=0.
        END DO
        ID(I,I)=ONE
      END DO
      ID(NISS,NISS)=ONE
C*** IDENTITY VECTOR
      DO I=1,NISS
        IF(I.LE.3)AS(I)=ONE
        IF(I.GE.4)AS(I)=0.
      END DO
C*** LAME COEFF.
      XMU=E0/(ONE+XNU)/TWO
      LAMDA=E0*XNU/(ONE-TWO*XNU)/(ONE+XNU)
C
      SMR=STATEV(1)
      SMP=STATEV(2)
      D=STATEV(3)
      DO I=1,NISS
        PSTRN(I)=STATEV(I+4)
      END DO
```

```
C*** TOTAL STRAIN
      DO I=1,NISS
        STRN(I)=STRAN(I)+DSTRAN(I)
        IF(I.GE.4)STRN(I)=STRN(I)*RAC2
      END DO
C*** ELASTIC PREDICTOR
      DSMP=0.
      IF((UNIAXI.EQ.'y').OR.(UNIAXI.EQ.'Y'))THEN
        STRN(2)=-XNU*STRN(1)-(0.5-XNU)*PSTRN(1)
        STRN(3)=STRN(2)
      END IF
C....Trace of strain tensor
      TREPS=STRN(1)+STRN(2)+STRN(3)
C....Effective stress
      DO I=1,NISS
        ESTRS(I)=LAMDA*TREPS*AS(I)+2.*XMU*(STRN(I)-PSTRN(I))
      END DO
      CALL INVAR(ESTRS,ESTRSH,ESTRSB,NISS)
      SIGN=1.
      IF(ESTRS(1).LT.0.)SIGN=-1.
C*** TEST THE YIELD CONDITION
      F=ESTRSB-SIGS
      IF(F.LE.0.)THEN
        GO TO 500
      ELSE
C*** IF THE ELASTIC PREDICTOR DOES NOT VERIFY THE YIELD CONDITION
        DO I=1,NISS
          N(I)=NN*(ESTRS(I)-ESTRSH*AS(I))/ESTRSB
        END DO
C*** Corrections over the elastic predictor
C....Correction over p
        DENOM=3.*XMU
        CSMP=F/DENOM
C....Correction over effective stress
        DO I=1,NISS
          CESTRS(I)=-2.*F*N(I)/3.
        END DO
C*** LOOP ON THE PLASTIC CORRECTIONS
        NFOIS=0
        CRITERE=.TRUE.
        DO WHILE(CRITERE)
          NFOIS=NFOIS+1
          DIV1=ESTRSB
          NTEST=2
          IF(NFOIS.EQ.1)NTEST=1
C*** UPDATE THE STATE
C....Update inc. of p
          DSMP=DSMP+CSMP
C....Update effective stress
          DO I=1,NISS
            ESTRS(I)=ESTRS(I)+CESTRS(I)
          END DO
          CALL INVAR(ESTRS,ESTRSH,ESTRSB,NISS)
          DO I=1,NISS
            N(I)=NN*(ESTRS(I)-ESTRSH*AS(I))/ESTRSB
          END DO
C
          IF(NTEST.EQ.2)THEN
            DIV2=ESTRSB
            REL=(DIV2-DIV1)/DIV1
          END IF
C....If we converge too slowly,or we diverge,exit.The main
C....program will propose a smaller "time" increment.
          IF(NFOIS.EQ.50.OR.
     &      (NTEST.EQ.2.AND.REL.GE.15.D-02))THEN
            CONVER='n'
            GO TO 500
          END IF
```

```
C*** Compute residual functions
          F=ESTRSB-SIGS
          IF((UNIAXI.EQ.'y').OR.(UNIAXI.EQ.'Y'))THEN
            STRN(2)=-0.5*STRN(1)+(0.5-XNU)*SIGN*SIGS/E0
            STRN(3)=STRN(2)
            TREPS=STRN(1)+STRN(2)+STRN(3)
          END IF
          DO I=1,NISS
            KESTRS(I)=ESTRS(I)-LAMDA*TREPS*AS(I)-
     &        2.*XMU*(STRN(I)-PSTRN(I)-DSMP*N(I))
          END DO
C....Max. residuals
          RESTRS=0.
          DO I=1,NISS
            IF(DABS(KESTRS(I)).GT.RESTRS)RESTRS=DABS(KESTRS(I))
          END DO
C*** ITERATIVE TEST ON THE YIELD CONDITION
          IF(DABS(F).LT.TOLRES.AND.RESTRS.LT.TOLRES)THEN
            CRITERE=.FALSE.
          ELSE
C*** PLASTIC CORRECTIONS
            CALL VTRANV1(N,KESTRS,XNKES,NISS)
C....Correction over p
            DENOM=3.*XMU
            CSMP=(F-XNKES)/DENOM
C....Correction over effective stress
            DENOM=DENOM*DSMP/ESTRSB+(1.)
            DO I=1,NISS
              CESTRS(I)=(2./3.)*(XNKES-DENOM*F)*N(I)-KESTRS(I)
              CESTRS(I)=CESTRS(I)/DENOM
            END DO
C*** END OF THE ITERATIVE TEST ON THE YIELD CONDITION
          END IF
C*** END OF THE LOOP ON THE PLASTIC CORRECTIONS
        END DO
C*** END OF TEST ON THE YIELD CONDITION
      END IF
500   CONTINUE
      IF(CONVER.EQ.'y')THEN
        IF((UNIAXI.EQ.'y').OR.(UNIAXI.EQ.'Y'))THEN
          DSTRAN(2)=STRN(2)-STRAN(2)
          DSTRAN(3)=DSTRAN(2)
        END IF
C....Plastic strain increment
        DO I=1,NISS
          DPSTRN(I)=N(I)*DSMP
        END DO
C....Damage increment
        DD=0.
        CALL DAMAGE(TREPS,ESTRSB,Y,RNU)
        IF(DSMP.GT.0..AND.SMP+DSMP.GE.SMPD.AND.
          COUPL.EQ.'y')THEN
          DD=Y*DSMP/S0
        END IF
        STAR=(ONE-D-DD)*DSQRT(2.*E0*Y)
C....Plastic multiplier increment
        DSMR=(ONE-D-DD)*DSMP
C....Compute stresses
        DO I=1,NISS
          STRS(I)=ESTRS(I)*(ONE-D-DD)
        END DO
        STRSB=ESTRSB*(ONE-D-DD)
C*** STORE STATEV AT THE END OF THE INCREMENT
        STATEV(1)=SMR+DSMR
        STATEV(2)=SMP+DSMP
        STATEV(3)=D+DD
        STATEV(4)=DD
        DO I=1,NISS
          STATEV(I+4)=PSTRN(I)+DPSTRN(I)
        END DO
```

```
        STATEV(11)=DSMP
      END IF
      RETURN
      END
C************************************
      SUBROUTINE INVAR(V,VH,VB,NISS)
C************************************
C... 1st and 2nd stress invariants
      IMPLICIT REAL*8(A-H,O-Z)
      REAL*8 V(6)
      COMMON/ETIQ2/ONE,TWO,THREE
C
      VH=(V(1)+V(2)+V(3))/THREE
      CONST=0.
      DO I=1,NISS
        IF(I.LE.3)CONST=CONST+(V(I)-VH)*(V(I)-VH)
        IF(I.GE.4)CONST=CONST+V(I)*V(I)
      END DO
      VB=DSQRT(THREE*CONST/TWO)
      RETURN
      END
C*******************************************
      SUBROUTINE DAMAGE(TREPS,ESTRSB,Y,RNU)
C*******************************************
      IMPLICIT REAL*8(A-H,O-Z)
      COMMON/ETIQ1/E0,XNU,SIGS,SMPD,S0
C
      TERM1=ESTRSB*ESTRSB*(1.+XNU)/E0
      TERM2=E0*TREPS*TREPS/(1.-2.*XNU)/2.
      Y=(TERM1+TERM2)/3.
      RNU=2.*E0*Y/ESTRSB/ESTRSB
      RETURN
      END
C***************************************
      SUBROUTINE VTRANV1(V,V1,VTV1,NISS)
C***************************************
C... Inner product of 2 symmetric 2nd order tensors
      IMPLICIT REAL*8 (A-H,O-Z)
      REAL*8 V(6),V1(6)
C
      VTV1=0.
      DO I=1,NISS
        VTV1=VTV1+V(I)*V1(I)
      END DO
      RETURN
      END
C*****************************************************
      SUBROUTINE OUTPUT(NU1,NU2,TIME,STRAN,STRS,STATEV,
     &  NTENS,NSTATV,STRSB,STAR)
C*****************************************************
      IMPLICIT REAL*8(A-H,O-Z)
      REAL*8 STRAN(NTENS),STRS(NTENS),STATEV(NSTATV)
C
      WRITE(NU1,130)TIME,(STRAN(I),I=1,3),(STRS(I),I=1,3),
     &               STATEV(3),STATEV(2),STRSB,STAR
      WRITE(NU2,130)TIME,(STRAN(I),I=4,6),(STRS(I),I=4,6),
     &               STATEV(3),STATEV(2),STRSB,STAR
      WRITE(90,444)TIME,STATEV(3),STATEV(2),STRSB,STAR
 130  FORMAT(1X,E10.4,' , ',3(E10.4,','),2X,3(E10.4,','),
             3X,E10.4,' , ',E10.4,' ,',2(E10.4,',') )
 444  FORMAT(3X,E10.4,5X,E10.4,3X,E10.4,3X,E10.4,4X,E10.4)
      RETURN
      END
```

4.3.4 Exercises Using the DAMAGE 90 Postprocessor

DAMAGE 90 may become your friend if you bring it to life in your computer. The following examples which were already performed with DAMAGE 90 will help you to set up the program without too many "bugs". Please try them.

Case of ductile damage

The material properties are those of a stainless steel:

$E = 200\,000$ MPa	$\nu = 0.3$	
$\sigma_f = 200$ MPa	$\sigma_y = 300$ MPa	$\sigma_u = 500$ MPa
$S = 0.06$ MPa	$\varepsilon_{pD} = 10\%$	$D_{1c} = 0.99$

Pure tension at microscale

Take as the main input

$p_D = 10\%$, $D_c = 0.99$ and ε_{11}. DAMAGE 90 computes the other strains:

$$\varepsilon_{22} = \varepsilon_{33} = -\frac{1}{2}\varepsilon_{11} + \left(\frac{1}{2} - \nu\right)\frac{\sigma_{11}}{E(1-D)}.$$

For a better representation of the large strain hardening, use the piecewise perfect plasticity procedure with

ε_{11} %	0	0.25	1.5	5	
σ_s MPa	200	300	400	500.	

The results are plotted in Figure 4.18.

Plane strain at meso-(or micro-)scale

Take the same material data as previously but use the perfect plasticity procedure with

$$\sigma_s = 500 \text{ MPa}.$$

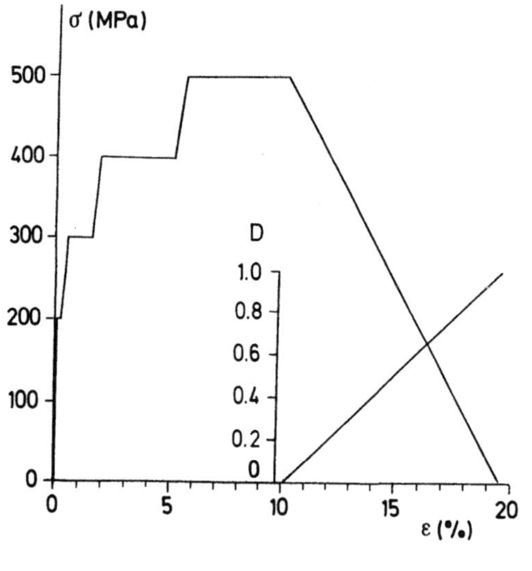

Fig. 4.18. Numerical simulation of a pure tension test at the microscale

Perform proportional loading in strain histories:

$$\varepsilon_{11} \text{ given}, \quad \varepsilon_{22} = \alpha\varepsilon_{11}, \quad \varepsilon_{33} = 0,$$

and find for each numerical test the set of strains $(\varepsilon_{11}, \varepsilon_{22})$ which corresponds to crack initiation. You will obtain the so-called ''damage forming limit curve'' in plane strain, as shown in Figure 4.19.

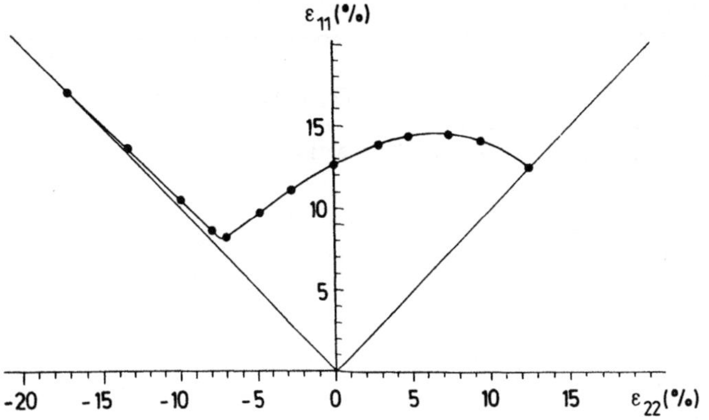

Fig. 4.19. Crack initiation limit curve in plane strain

Case of brittle damage

The material data are those of a ceramic:

E= 400 000 MPa	ν= 0.2	
σ_f= 200 MPa	σ_y= 250 MPa	σ_u= 300 MPa
S= 0.00012 MPa	ε_{pD}= 0	D_{1c}= 0.05

Perform calculations for the case of an elastic tension case at the mesoscale: ε_{11} given with $\varepsilon_{22} = \varepsilon_{33} = -\nu\varepsilon_{11}$, with a plastic threshold $\sigma_s = 200$ MPa, and plot the result at the mesoscale (an elastic-brittle material) and at the microscale, as in Figures 4.20 and 4.21.

Low and high cycle fatigue

The main hypothesis of the damage model is that there is no difference in the equations for low or high cycle fatigue because they obey the same energetic mechanism. Furthermore, written as a damage rate the constitutive equation is valid even if a cycle cannot be defined.

The material data is that of an aluminum alloy:

E= 72 000 MPa	ν= 0.32	
σ_f= 303 MPa	σ_y= 306 MPa	σ_u= 500 MPa
S= 6 MPa	ε_{pD}= 10%	D_{1c}= 0.99

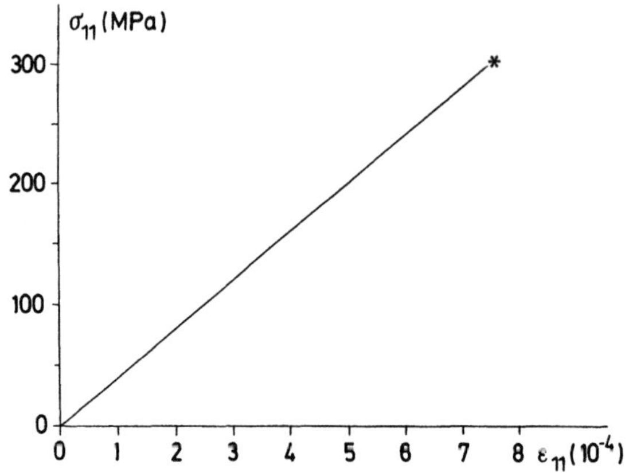

Fig. 4.20. Stress strain and brittle damage at the mesoscale in pure tension

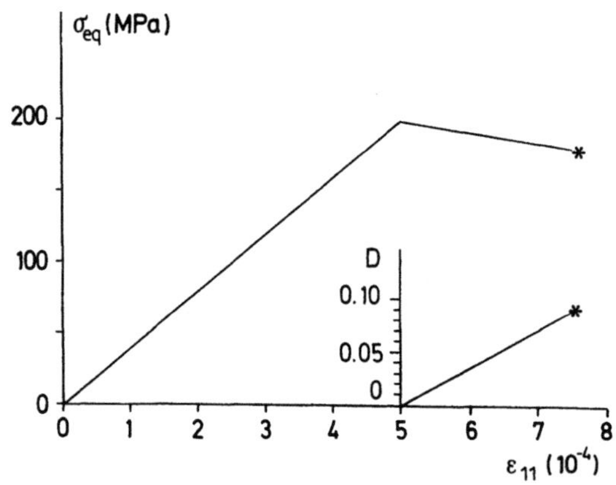

Fig. 4.21. Stress strain and damage at the microscale when pure tension is applied at the mesoscale

Consider a pure tension case at the microscale in which ε_{11} is the main input, the other components ε_{22} and ε_{33} being calculated by DAMAGE 90 as

$$\varepsilon_{22} = \varepsilon_{33} = -\frac{1}{2}\varepsilon_{11} + \left(\frac{1}{2} - \nu\right)\frac{\sigma_{11}}{E(1-D)}.$$

The cyclic strain hardening is taken into account by using piecewise perfect plasticity according to the cyclic stress strain curve:

ε_{11} %	$\pm$ 0.425	$\pm$ 0.43	$\pm$ 0.47	$\pm$ 1	$\pm$ 3.5	$\pm$ 4.5
σ_s MPa	300	305	308	370	440	460.

Perform some numerical simulations to obtain the number of cycles to crack initiation as a function of the stress amplitude, and plot the results together with the number of cycles N_0 to crack nucleation corresponding to $p = p_D$ (Figure 4.22). Consider the case $\varepsilon_{11} = \pm 3.5\%$ ($\Delta\varepsilon = 7\%$). The number of cycles to failure is 40. Compare to an hypothetical case of pure tension at the mesoscale where the input is

$$\varepsilon_{11} = \pm 3.510^{-2}, \quad \varepsilon_{22} = \varepsilon_{33} = -\nu\varepsilon_{11}.$$

Due to the effect of the triaxiality the number of cycles to failure would be only 8. Plot the graphs of stresses, strains and damage for the pure tension case at the microscale as in Figure 4.23.

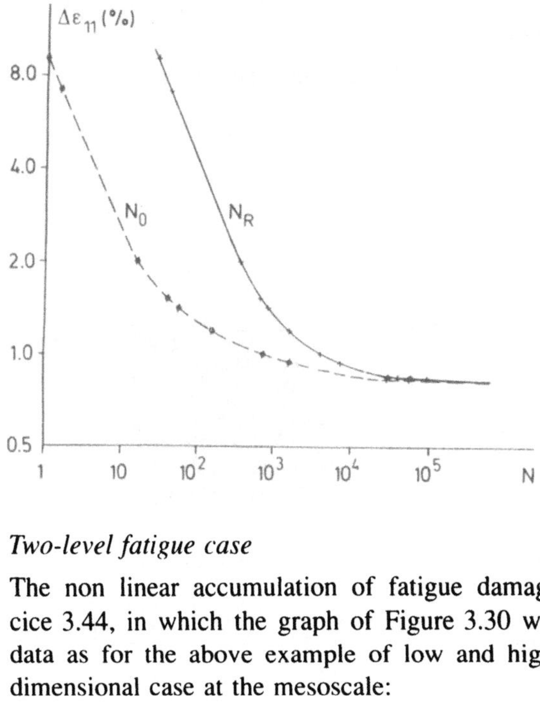

Fig. 4.22. Fatigue rupture curve

Two-level fatigue case

The non linear accumulation of fatigue damage was also dealt with in Exercice 3.44, in which the graph of Figure 3.30 was drawn with the same material data as for the above example of low and high cycle fatigue and for the one-dimensional case at the mesoscale:

$$\varepsilon_{22} = \varepsilon_{33} = -\nu\varepsilon_{11}, \quad \varepsilon_{12} = \varepsilon_{13} = \varepsilon_{23} = 0.$$
$$\varepsilon_{11} = \pm 0.47\% \rightarrow N_R = 7720 \text{ cycles;}$$
$$\varepsilon_{11} = \pm 0.425\% \rightarrow N_R = 109570 \text{ cycles;}$$

$\varepsilon_{11} = \pm 0.47\%$ during 3088 cycles and $\varepsilon_{11} = +0.425\%$ after $\rightarrow N_R \simeq 52400$ cycles;
$\varepsilon_{11} = 0.425\%$ during 43828 cycles and $\varepsilon_{11} = +0.47\%$ after $\rightarrow N_R \simeq 50000$ cycles;
Check that for

— the highest loading first, $\dfrac{N_1}{N_{R1}} + \dfrac{N_2}{N_{R2}} = 0.85 < 1;$

— the lowest loading first, $\dfrac{N_1}{N_{R1}} + \dfrac{N_2}{N_{R2}} = 1.2 > 1.$

Figure 4.24 gives the results $\frac{N_2}{N_{R2}}$ when $\frac{N_1}{N_{R1}}$ varies in both cases. Compare these with Figure 3.31.

Multiaxial fatigue

You may also play with DAMAGE 90 to obtain contours of the number of cycles to failure in biaxial fatigue as it is shown for the same aluminum alloy as previously:

— for plane strain in Figure 4.25:

$$\varepsilon_{11} = \pm x, \quad \varepsilon_{22} = \pm y, \quad \varepsilon_{33} = 0;$$

— for tension and shear in Figure 4.26:

$$\varepsilon_{11} = \pm x, \quad \varepsilon_{12} = \pm y, \text{ all other components } - 0;$$

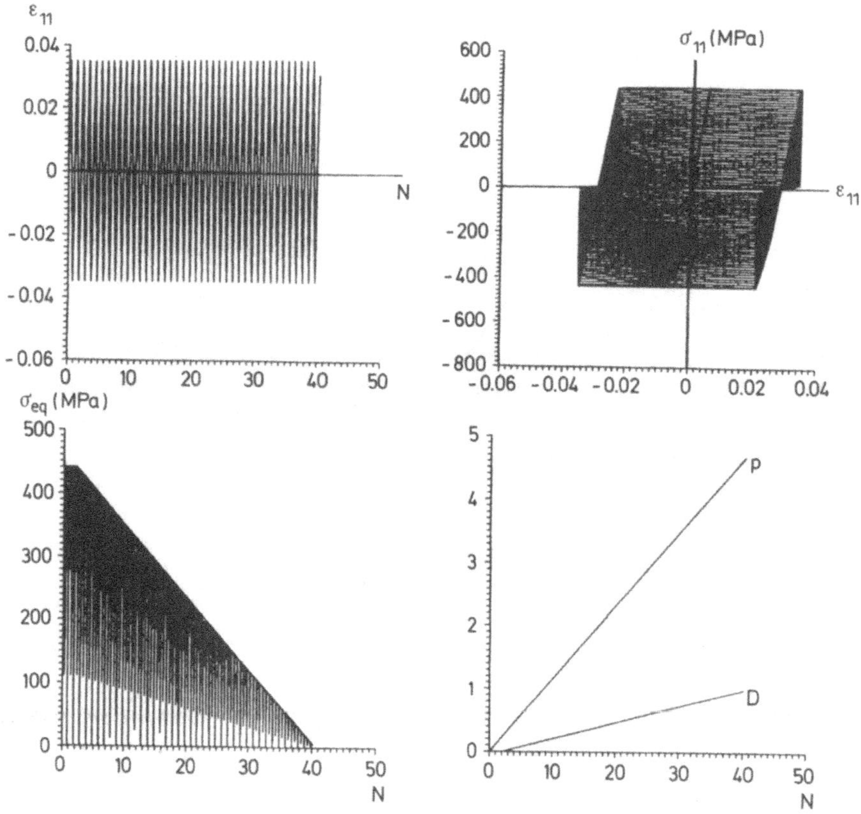

Fig. 4.23. Very low cycle fatigue simulation in tension at the microscale

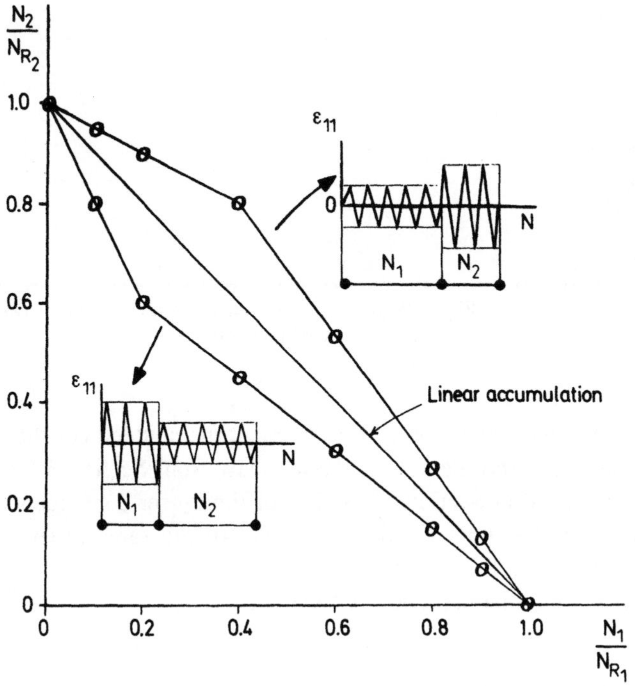

Fig. 4.24. Accumulation diagram for two-level simulation in tension at the mesoscale

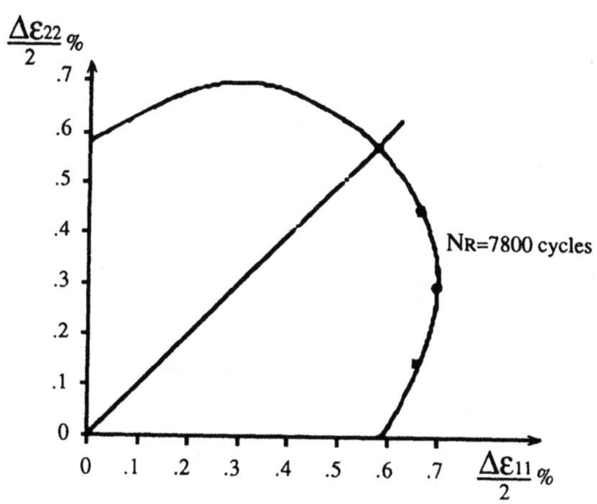

Fig. 4.25. Biaxial fatigue envelope for the case of tensile plane strains imposed

4.4 Fully Coupled Analysis

For ductile-and creep-type damage, the damage is often distributed over a large volume of the structural component under consideration. Then, for a good accuracy the coupling between damage and strain must be considered all over the structure or at least over a substructure. This makes the calculation (corresponding to the scheme in Figure 4.27) more complex, but there is no other way to proceed.

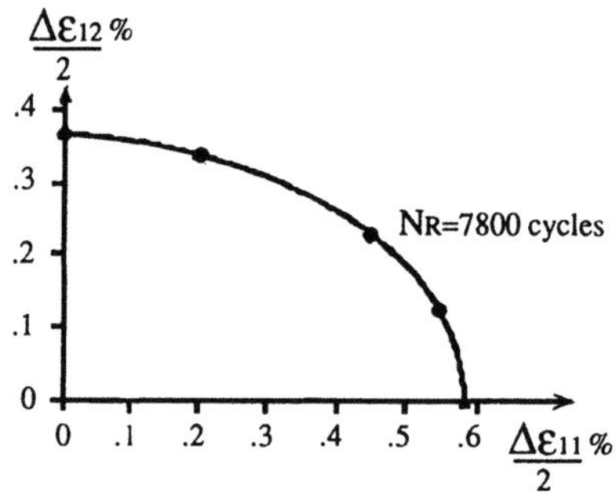

Fig. 4.26. Biaxial fatigue envelope for the case of tensile and shear strains imposed

The classical way to perform this calculation is to implement the coupled constitutive equations from Section 2.1.4 and the kinetic damage law from Section 3.1 in a finite element code as explained in Section 4.1.3. The difficulty for convergence of iterations and for the computer time of calçulations are of the same order as for classical elastoplastic analysis. Such routine is available in the French code **SYSTUS** and possible with the American code **ABAQUS**.

4.4.1 Initial Strain Hardening and Damage

What makes this method interesting is that it allows us to take into account pre-existing strain hardening and damage as initial fields in some part of the structural component to be analyzed. It is always of great interest to evaluate the remaining strenght or lifetime of a component after an accident which has induced plastic deformation and damage.

Three quantities may be introduced as material initial conditions at the Gauss points of the finite element meshes:

— R_0, the isotropic strain hardening stress as an initial value for its differential constitutive equation,

$$\dot{R} = b(R_\infty - R)\dot{\lambda},$$

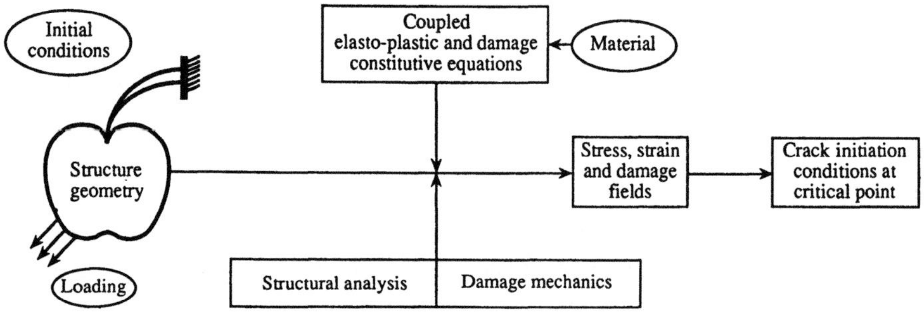

Fig. 4.27. Fully coupled analysis of strain and damage

or, in the absence of damage, p_0, an initial value of the accumulated plastic strain since in this case, there is a one-to-one relationship between p and R:

$$R = R_\infty[1 - \exp(-bp)];$$

— $\mathbf{X}_0^D$, the kinematic strain hardening stress tensor as an initial value for its differential constitutive equation,

$$\dot{X}_{ij}^D = \gamma\left[\frac{2}{3}X_\infty\dot{\varepsilon}_{ij}^p(1-D) - X_{ij}^D\dot{\lambda}\right];$$

— D_0 an initial value of damage for its kinetic law,

$$\dot{D} = \frac{Y}{S}\dot{p}.$$

One more quantity which often exist as a characteristic of the structure, but not of the material, is a state of residual stresses, after welding, for example. A self-equilibrated residual stress field can also be introduced at the Gauss points as an initial condition:

$$\sigma_{ij}^0 \text{ such that } \left\{\begin{aligned} \sigma_{ij}^0{}_{,j} &= 0 \text{ in } S \\ \sigma_{ij}^0 n_j &= 0 \text{ on } \partial S \end{aligned}\right\} \text{ Figure 4.1}$$

The first three material quantities are not difficult to introduce in a finite element code, but the main problem is to identify them. As far as isotropic strain hardening and damage are concerned, if the damaged structure is available for testing, one can evaluate R_0 and D_0, at least on the surface, by microhardeness measurements as explained in Section 1.3.3. As far as kinematic strain hardening is concerned its initial value may be evaluated by means of phenomenological considerations(!!) or neglected; the latter is not an approximation for the stabilized cycle which does not depend upon the back stress $\mathbf{X}_0^D$:
Note that those initial conditions may be considered in the uncoupled analysis as a constant reduction of the elasticity modulus and the strain hardening variables.

4.4.2 Example of a Calculation Using the Finite Element Method (R. BILLARDON, 1988)

It is out of the scope of this course to describe in detail the finite element procedure for elastoplasticity or viscoplasticity coupled with damage calculations. The set of constitutive equations of Section 2.1.4 together with the kinetic damage law identified in Section 3.1.4 may be implemented in any classical finite element code having plasticity routines. Best results are obtained with the algorithm described briefly in Section 1.3 of the present chapter.
The practical case considered here schematically is an underground gas pipeline which has been damaged by the shock of a machine digging in the soil. The pressure inside the pipe varies periodicallly twice a day due to gas consumption. It induces fatigue of the material especially where the shock occured. The question is: What is the remaining strength or lifetime of the damaged pipe before it breaks? Obviously, it is of primary importance to decide when and how to make a repair.

The pipe and the loading being considered are defined in Figure 4.28.
A simple calculation of the undamaged cylinder unconstrained at the ends by the expansion joints gives the state of stress everywhere. Assuming the material to be elastic, one can deduce the displacements everywhere and particularly at the boundaries A and B of the substructure containing the damage that will be studied in more detail. Figure 4.29 shows the geometrical model submitted to a one-dimensional displacement within the hypothesis of plane strain. Note that this substructuring neglects the redistribution of stress due to plasticity that will occur at the root of the notch as a result of the loading $\vec{u}_{(t)}$.
The material is a low-carbon steel. Its elastoplastic and damage parameters have been determined from several tension tests according to procedures described in Sects.2.1.5 and 3.1.5:

— E, ν for elasticity;
— σ_y, R_∞ b, X_∞, γ for plasticity;
— S, ε_{p_D}, D_{1c} for damage together with σ_f and σ_u.

More interesting is the strain hardening and the damage induced by the shock in the idented zone; these must be considered as initial conditions, as explained in Section 4.4.1. These damage and strain hardening fields have been identified from microhardness experiments on a polished section of a plate being subjected to a shock similar to that which damaged the pipe. The method used is described in Section 1.3.3. The results are shown in Figure 4.30.

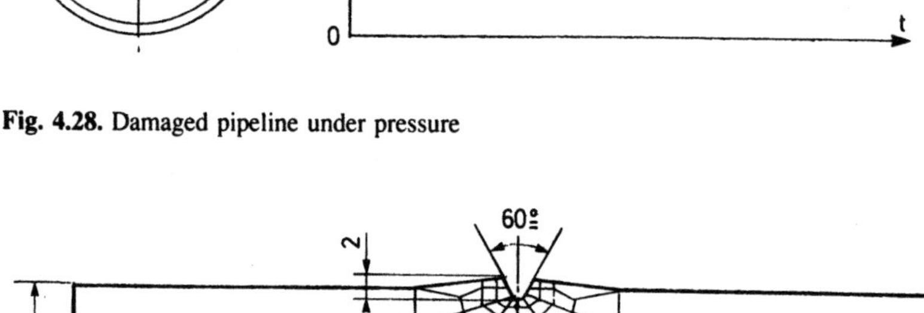

Fig. 4.28. Damaged pipeline under pressure

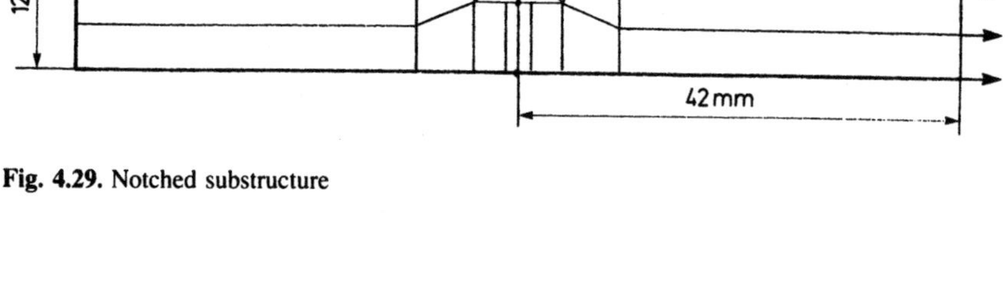

Fig. 4.29. Notched substructure

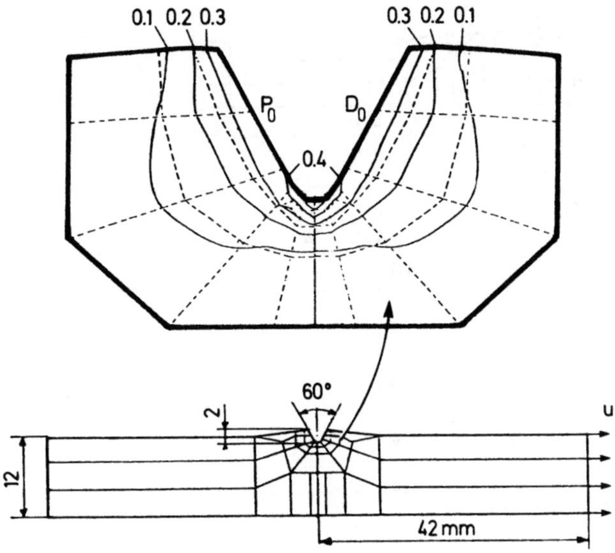

Fig. 4.30. Initial strain-hardening and damage fields

The values of R_0 and D_0 are introduced at each Gauss point of the finite element meshes represented in Figure 4.30. The rest of the meshing is indicated in Figure 4.29. The results of the structural calculation corresponding to a monotonic loading performed in order to determine the remaining strength, are shown in Figures 4.31 and 4.32.

From the evolution of the damage in Figure 4.31, the displacement which will cause crack initiation is $u \simeq 0.12$ mm. It turns out to be about 50% of the elastic displacement corresponding to the load of the pipe when the von Mises equivalent stress is equal to the yield stress σ_y. This particular damage has reduced the strenght of the pipe by a factor of about 50%.

If the initial strain hardening and damage are not taken into account the damage evolution shows a critical displacement of about 0.20 mm, that is, an error of 66%.

As far as crack initiation by fatigue due to the cyclic pressure in the pipe is concerned, the same calculation is performed with a periodic displacement as input. To avoid a considerable amount of calculation, the procedure of integration by damage increments (instead of time or cycle increments) described in Section 4.13 is used. For the case under consideration, the number of cycles to crack initiation of the indented substructure (or the pipe) was about 100 times less than the fatigue life of the initial pipe. As the nominal number of cycles to failure of the undamaged pipe line was of the order of 10^5, this corresponds to 1000 cycles, that is, more than a year to organize the repair without cutting off gas delivery. Thank you.

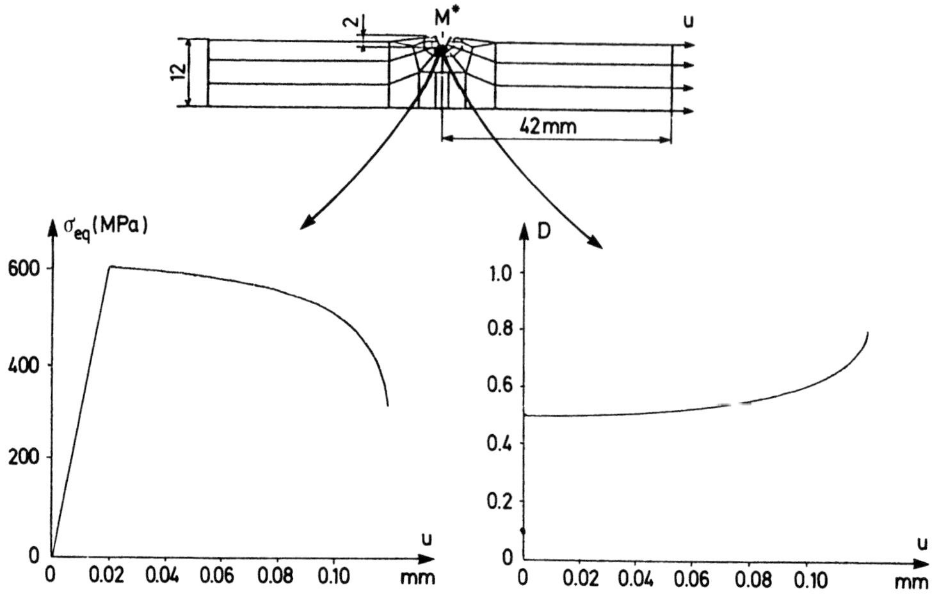

Fig. 4.31. Evolution of the von Mises equivalent stress and the damage at the critical point M^*

4.4.3 Growth of Damaged Zones and Macrocracks

Damage mechanics may predict crack initiation at the mesoscale, generally with a damaged zone surrounding the cracked mesovolume element. As the loading, or the time, continue to increase, the calculation may continue with a fully damage point until another mesovolume element reachs its critical state of damage, and so on, to describe the evolution of a fully damaged zone. No stress intensity factors, no fracture mechanics to model the phenomenon of crack growth. Nice isn't it! But nice only in principle because this ideal process does not work so easily, due to the localization of damage at its critical stage. The problem is no longer ellptic (see Section 2.3.5). A localized bifurcation of the solution occurs and, if classical finite elements are used, the convergence regarding the mesh size is not ensured.

We may say in 1991 that this is still an open problem despite several attempts to introduce concepts which avoid this difficulty:

— neglecting the coupling between damage and elasticity: D. Hayhurst (1978); J.L. Chaboche (1982);

— partial regularization of the problem by viscosity: A. Needleman (1988);

— introduction of a characteristic length for each material taken as the minimum size of the meshes in the finite element calculation: Z. Bazant (1976);

— definition of a ''nonlocal'' damage variable as a weighted mean value over a characteristic volume: G. Pijaudier-Cabot and Z. Bazant (1986), K. Saanouni (1988);

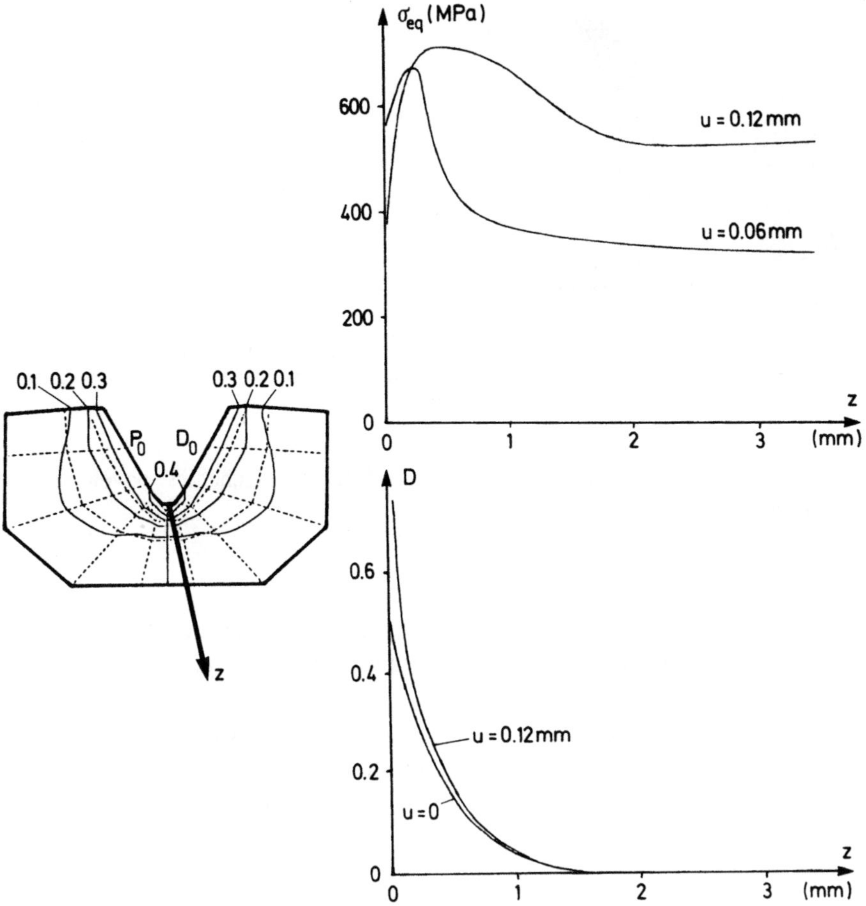

Fig. 4.32. Von Mises equivalent stress and damage along the z axis

— derivation of special finite elements able to model discontinuities of strains due to damage localization: M. Ortiz (1987), T. Belyschko (1988), R. Billardon and I. Doghri, (1989);

Then, to cover the whole process of fracture of a component, the following steps may be worked out (Figure 4.33):

1) structural calculations either as an uncoupled, locally coupled or fully coupled analysis,
2) conditions of mesocrack initiation:
   * loading or time or number of cycles;
   * direction $\vec{n}$ of the plane of the crack determined by $\det(\vec{n} \cdot \mathbf{H} \cdot \vec{n}) = 0$ in elastoplasticity (see Section 2.3.5);
   * crack area: its value may be derived from the concept of characteristic

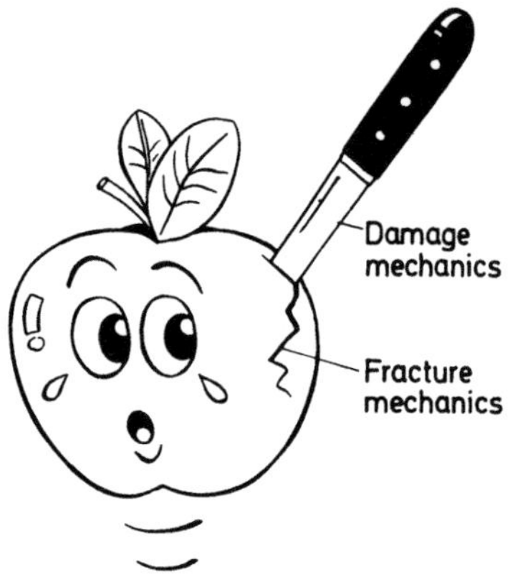

Fig. 4.33. "The frightened apple": damage mechanics and fracture mechanics as complementary sciences (after J- P. Balloffet)

length matching damage mechanics and fracture mechanics developed in Section 2.3.1.

$$l = \frac{G_c}{Y_c}$$

for a two-dimensional problem, where l is the crack length of the mesocrack initiated by the damage process. For a three-dementional problem, take l^2 as the crack area.

* the crack shape is not a problem in two dimensions. In three dimensions, a circular crack can be asumed, as it is the shape for which the stress intensity factor in mode I is constant all along the circumference;

3) Fracture mechanics analysis to obtain the growth of the initial crack as defined above.

4.4.4 Exercise on Damaged Zone at a Crack Tip

Considering a macrocrack in a two-dimensional structure made of a ductile material and loaded in mode I, we may consider the stress field as elastic far away from the crack tip then elastoplastic near the crack tip and much affected by damage in the process zone very close to the crack tip. Let us try to analyze some features of this process zone (Figure 4.34).

Assuming:
— small deformations,
— small-scale yielding,
— a strain-hardening power law

$$p = \left(\frac{\sigma_{eq}}{K_p}\right)^{M},$$

which induces a stress field with the H.R.R. singularity (Hutchinson-Rice and Rosengreen) in polar coordination (r, θ):

$$\sigma_{eq} \sim \frac{1}{r^{\frac{1}{M+1}}},$$

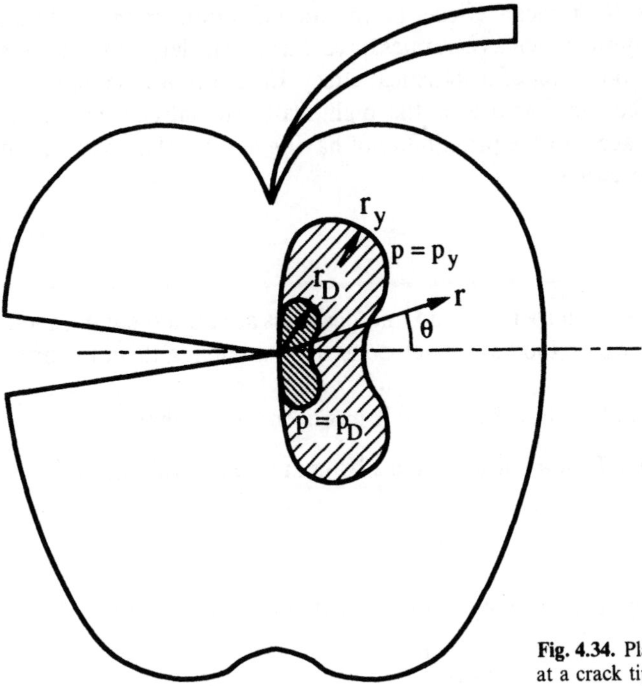

Fig. 4.34. Plastic zone and damaged zone at a crack tip

— stress and strain redistribution due to damage negligible,
— the plastic zone $r_Y(\theta)$ delimited from the elastic zone by an offset $p_Y (p_Y \approx 0.2\ 10^{-2}$ for example),
— the damaged zone $r_D(\theta)$ delimited from the plastic zone by the damage threshold p_D.

Show that the damaged zone may be deduced from the plastic zone by an homotetic transformation of ratio

$$\frac{r_D}{r_Y} = \left(\frac{p_Y}{p_D}\right)^{\frac{M+1}{M}}.$$

For most of materials this damaged zone is very small.

Example: $p_Y = 0.2\ 10^{-2}$, $p_D = 10\ 10^{-2}$, $M = 5$, $\frac{r_D}{r_Y} \approx 10^{-2}$.

As the energies involved are more or less proportional to r_D^2 and r_Y^2, this corresponds to a very small energy dissipated in damage and fracture in comparison to the energy dissipated in plasticity. This is the reason why the critical value of the strain energy release rate G_C at fracture is so small in comparison to the strain energy involved.

4.5 Statistical Analysis with Microdefects (F. Hild 1994)

The fracture of any material is mainly governed by defects at a micro level which always exist after the process of elaboration. In the deterministic analysis, they are taken into account by their mean effects in the identification of the material parameters. But for materials such as ceramics or cast iron. The large size of some defects imposes the introduction of a statistical effect. Furthermore the only way to explain the size effect on fracture or the high scatter in fatigue failure for example is to take into account the probability of having defects of a given size in the tested sample or structure.

4.5.1 Initial defects

Since the geometries of the microcracks and microcavities are always complex and most of the time unknown, a good approach consists in modeling them by initial equivalent damages.
Let us recall the results of Section 2.3 where it has been shown that:

— the equivalent damage D of a crack of area s lying in a cell of volume $l \times l \times l$ is

$$D \approx \frac{s}{l^2}$$

— the equivalent damage D of a cavity of diameter d in a cell of volume $l \times l \times l$ is

$$\mathrm{D} \approx \frac{d^2}{l^2}.$$

Then, the statistical characteristics of the surface of initial defects may be directly translated into the statistics of initial damage D_0 provided the definition of the size a representative volume element be possible. Practically, this size l may be the mean distance between the centers of adjacent microcracks, that is the inverse of their linear density. These defects may be randomly distributed in size, orientation, shape ... but we restrict ourselves to the distribution in size only given by the probability density $f(D_0)$ of the equivalent initial damage D_0

$$\operatorname{Prob}(D_0 \le D_0 \le D_0 + dD_0) = f(D_0)\, dD_0 .$$

The function $f(D_0)$ may be deduced from measurements of defects through observations with a microscope. We shall see that some statistical characteristics may be obtained more easily from mechanical characteristics at measoscale.
The probability density characterizes the distribution of the defects or the initial damage in a structure but it is also the probability of having an initial damage of a given value in each representative volume element.

4.5.2 Case of Brittle Materials

This case applies to ceramics or cast iron. Consider a Representative Volume Element damaged by one defect characterized by an equivalent initial damage D_0 having a probability density $f(D_0)$ and loaded by a damage equivalent stress σ^*. We look for statistics on the stress σ^*_R which induces a crack initiation, that is when the pure brittle failure criterion is reached. From Section 3.2.1 this is when

$$Y = Y_C = \frac{\sigma_u^2}{2E},$$

σ_u being the ultimate stress of the material. With

$$Y = \frac{\sigma^{*2}}{2E(1-D_0)^2},$$

$$\sigma_R^* = \sigma_u(1-D_0),$$

D_0 being a random variable $0 \le D_0 < 1$, the mean value of the failure stress, also a random variable, is

$$\bar{\sigma}_R^* = \int_0^1 \sigma_u(1-D_0)\, f(D_0)\, \mathrm{d}D_0,$$

$f(D_0)$ being a probability density

$$\int_0^1 f(D_0)\mathrm{d}D_0 = 1,$$

$$\bar{\sigma}_R^* = \sigma_u\left[1 - \int_0^1 D_0 f(D_0)\mathrm{d}D_0\right],$$

or

$$\boxed{\bar{\sigma}_R^* = \sigma_u(1-\bar{D}_0)},$$

$\bar{D}_0 = \int_0^1 D_0 f(D_0)\mathrm{d}D_0$ being the mean value of the initial damage distribution.

The standard deviation of the stress at fracture is obtained by

$$(\overline{\overline{\sigma_R^*}})^2 = \int_0^1 (\sigma_{R(D_0)}^* - \bar{\sigma}_R^*)^2 f(D_0)\mathrm{d}D_0,$$

$$(\overline{\overline{\sigma_R^*}})^2 = \int_0^1 [\sigma_u(1-D_0) - \sigma_u(1-\bar{D}_0)]^2 f(D_0)\mathrm{d}D_0,$$

$$(\overline{\overline{\sigma_R^*}})^2 = \sigma_u^2 \int_0^1 (\bar{D}_0 - D_0)^2 f(D_0)\, \mathrm{d}D_0,$$

$$\boxed{\overline{\overline{\sigma_R^*}} = \sigma_u \overline{\overline{D_0}}},$$

$\overline{\overline{D_0}} = \left[\int_0^1 (D_0 - \bar{D}_0)^2 f(D_0)\mathrm{d}D_0\right]^{1/2}$ being the standard deviation of the initial damage distribution.

As an example, consider an uniform distribution of initial damage

$$f(D_0) = \frac{1}{D_M} \quad 0 \le D_0 \le D_M \le 1,$$

the mean value and the standard deviation of the failure stress are

$$\bar{\sigma}_R^* = \sigma_u\left(1 - \int_0^{D_M} \frac{D_0}{D_M}\,dD_0\right) = \sigma_u\left(1 - \frac{D_M}{2}\right),$$

$$\bar{\bar{\sigma}}_R^* = \sigma_u\left[\int_0^{D_M}\left(D_0 - \frac{D_M}{2}\right)^2 \frac{1}{D_M}\,dD_0\right] = \frac{D_M}{2\sqrt{3}}.$$

4.5.3 Case of Quasi-Brittle Materials

Let us recall that materials considered as quasi-brittle are those for which plastic strain axists only at microscale without any substantial effect at mesoscale.
The same representative volume element damaged by one defect of magnitude D_0 as previously is considered but the damage may grow stable on reaching its critical value D_C corresponding to crack initiation. Section 3.2.2 gives the damage evolution law for quasi-brittle materials taken here with a zero value of the two thresholds for simplicity

$$\dot{D} = \frac{\sigma_f^2}{2ES} R_v \dot{\varepsilon}_{eq},$$

where σ_f is the fatigue limit of the material, S the damage strength, R_v the triaxiality function.
Assuming a proportional loading for which $R_v = \text{const}$.

$$D_{(t)} = D_0 + \frac{\sigma_f^2}{2ES} R_v \varepsilon_{eq(t)},$$

where ε_{eq} is the accumulated strain

$$\varepsilon_{eq} = \int_0^t \left(\frac{2}{3}\dot{\varepsilon}_{ij}^D \dot{\varepsilon}_{ij}^D\right)^{1/2} dt\,.$$

The time to crack initiation t_c is reached when $D = D_c$, considered here as a constant,

$$\varepsilon_{eq(t_c)} = \frac{2ES}{\sigma_f^2 R_v}(D_c - D_0) \quad \text{with } 0 \le D_0 < D_c$$

D_0 becoming a random variable with a probability density $f(D_0)$, the mean value of the accumulated stain to crack initiation is

$$\bar{\varepsilon}_{eq}^R = \frac{2ES}{\sigma_f^2 R_v}\int_0^{D_c}(D_c - D_0)\,f(D_0)\,dD_0\,,$$

or with

$$\int_0^{D_c} f(D_0)\,dD_0 = 1$$

$$\bar{\varepsilon}_{eq}^R = \frac{2ES}{\sigma_f^2 R_v}(D_c - \bar{D}_0)\,.$$

Its standard deviation is given by

$$\left(\overline{\overline{\varepsilon_{eq}^R}}\right)^2 = \int_0^{D_c} \left(\varepsilon_{eq(D_o)}^R - \bar{\varepsilon}_{eq}^R\right)^2 f(D_0)\, dD_0\,,$$

$$\left(\overline{\overline{\varepsilon_{eq}^R}}\right)^2 = \left(\frac{2ES}{\sigma_f^2 R_v}\right)^2 \int_0^{D_c} (\bar{D}_0 - D_0)^2 f(D_0)\, dD_0\,,$$

$$\boxed{\overline{\overline{\varepsilon_{eq}^R}} = \frac{2ES}{\sigma_f^2 R_v}\overline{\overline{D_0}}}\,,$$

where $\overline{\overline{D_0}} = \left[\int_0^{D_c} (D_0 - \bar{D}_0)^2 f(D_0)\, dD_0\right]^{1/2}$ is the standard deviation of the initial damage distribution.

— Application to uniaxial periodical fatigue of strain amplitude $\Delta\varepsilon \cdot (R_v = 1)$.

For this case $\varepsilon_{eq} = 2\Delta\varepsilon N$,
N being the number of cycles.

At crack initiation

$$\varepsilon_{eq}^R = 2\Delta\varepsilon N_R\,.$$

the mean value of the number of cycles to failure $\bar{N}_R$ and its standard deviation $\overline{\overline{N_R}}$ immediately follow

$$\bar{\varepsilon}_{eq}^R = 2\Delta\varepsilon \bar{N}_R \;\rightarrow\; \bar{N}_R = \frac{ES}{\sigma_f^2 \Delta\varepsilon}(D_c - \bar{D}_0)\,,$$

$$\overline{\overline{\varepsilon_{eq}^R}} = 2\Delta\varepsilon \bar{\bar{N}}_R \;\rightarrow\; \bar{\bar{N}}_R = \frac{ES}{\sigma_f^2 \Delta\varepsilon}\bar{\bar{D}}_0\,.$$

Writing the standard deviations as

$$\bar{\bar{N}}_R = \frac{\bar{N}_R}{D_c - \bar{D}_0}\bar{\bar{D}}_0$$

explains a well-known experimental result which shows that the standard deviation of the number of cycles to failure $\bar{\bar{N}}_R$, or the scatter in fatigue, increases with the number of cycles to failure $\bar{N}_R$. An example is given in Figure 4.35.

Another interesting consequence is related to the identification of initial damage or defects of a given material for which the Woehler curve is known by the mean value of the number of cycles to failure $\bar{N}_R$ and its standard deviation $\bar{\bar{N}}_R$. Solving the above equations for $\bar{D}_0$ and $\bar{\bar{D}}_0$ gives the mean value and the standard deviation of initial damage

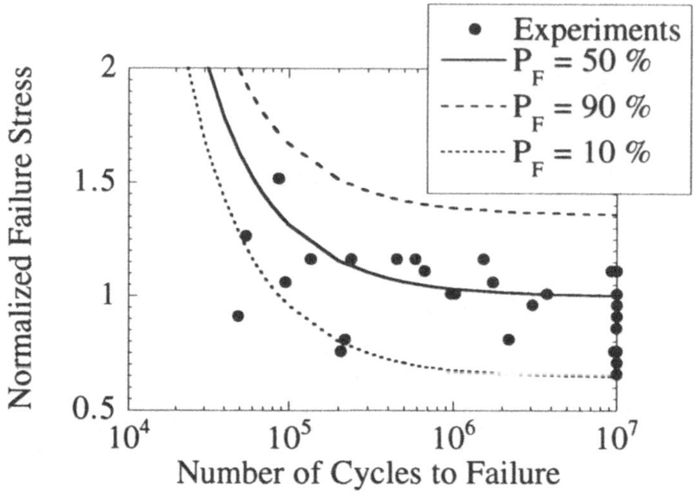

Fig. 4.35. Woehler curve of a SG cast iron with its scatter (from A.S. Beranger)

$$\boxed{\bar{D}_0 = D_c - \frac{\dot{\sigma}_f^2 \Delta\varepsilon}{ES} \bar{N}_R\,,}$$

$$\boxed{\bar{\bar{D}}_0 = \frac{\sigma_f^2 \Delta\varepsilon}{ES} \bar{\bar{N}}_R = \frac{D_c - \bar{D}_0}{\bar{N}_R} \bar{\bar{N}}_R\,.}$$

4.5.4 Case of Ductile Materials

The same kind of analysis may be performed with the damage evolution low of Section 3.3.1.

$$\dot{D} = \frac{\sigma_s^2 R_v}{2ES} \dot{p}\,,$$

or in the case of proportional loading with an initial damage D_0

$$D = D_0 + \frac{\sigma_s^2 R_v}{2ES} p\,.$$

When the value of the initial damage is randomly distributed with a mean value $\bar{D}_0$ und u standard deviation $\bar{\bar{D}}_0$, the mean value $\bar{P}_R$ and the standard deviation $\bar{\bar{P}}_R$ of the accumulated plastic strain at crack initiation defined by $D = D_c$ are

$$\boxed{\bar{P}_R = \frac{2ES}{\sigma_s^2 R_v} (D_c - \bar{D}_0)\,,}$$

$$\boxed{\bar{\bar{P}}_R = \frac{2ES}{\sigma_s^2 R_v} \bar{\bar{D}}_0 = \frac{\bar{P}_R}{D_c - \bar{D}_0} \bar{\bar{D}}_0\,.}$$

4.5.5 Volume Effect

It has been observed many times that the stress to failure of a specimen or a component decreases when the size of this component increases. This phenomenon is important in brittle materials and it is mainly due to the probability of finding a large defect which increases when the considered volume is larger.

Consider a structure of volume V loaded by a uniform stress field and having a distribution of initial damage $f(D_0)$.

Consider the representative volume elements of the material having a volume V_0, their number in the structure is

$$n = \frac{V}{V_0}.$$

The hypothesis of the weakest link is adopted in the sense that if a crack initiation occurs in an RVE, the whole structure breaks.

Let us call the failure of the structure probability P_F and the corresponding probability of survival $P_S = 1 - P_F$.

Considering the case of brittle materials and no interaction between defects, a crack initiation occurs in an RVE (i) if

$$\sigma_i^* \geq \sigma_u(1 - D_c) = \sigma_R.$$

The probability of survival of the structure considered as the set of the RVE links is given by the product

$$P_S = 1 - P_F = \prod_{i=1}^{n} P_{Si}(\sigma_i^* < \sigma_R).$$

Taking the Logarithm

$$Ln(1 - P_F) = \sum_{i=1}^{n} Ln(1 - P_{Fi}),$$

but

$$\frac{1}{n}\sum_{i=1}^{n} Ln(1 - P_{Fi}) \approx \frac{1}{V}\int_V Ln(1 - P_{F0})\,\mathrm{d}V,$$

where

$$P_{F0} = \int_{D_c}^{1} f(D_0)\,\mathrm{d}D_0,$$

then

$$Ln(1 - P_F) = \frac{1}{V_0}\int_V Ln(1 - P_{F0})\,\mathrm{d}V,$$

or

$$P_F = 1 - \exp\left(\frac{1}{V_0}\int_V Ln(1 - P_{F0})\,dV\right).$$

If the probability P_{F0} does not depend upon the space coordinates,

$$P_F = 1 - \exp\left(\frac{V}{V_0} Ln(1 - P_{F0})\right)$$

or

$$\boxed{P_F = 1 - (1 - P_{F0})^{V/V_0}}.$$

This formula explains the size effects: V_0 being fixed by the nature of the material, if the volume of the structure V increases, the probability of failure P_F also increases.
The well-known Weibull formula may be derived for a three-dimensional state of stress with two additional hypothesis.

— The probability P_{F0} is small with respect to 1,

$$Ln(1 - P_{F0}) \approx -P_{F0},$$

$$P_F = 1 - \exp\left(-\frac{V}{V_0} P_{F0}\right).$$

— The probability density of initial damage is a power-law function

$$f(D_0) = m(1 - D_0)^{m-1},$$

where m is the Weibull exponent

$$P_{F0} = \int_{D_c}^{1} f(D_0)\,\mathrm{d}D_0 = (1 - D_c)^m,$$

with
$$D_c = 1 - \frac{\sigma^*}{\sigma_u}.$$

The Weibull formula is

$$\boxed{P_F = 1 - \exp\left(-\frac{V}{V_0}\left[\frac{\sigma^*}{\sigma_u}\right]^m\right)}.$$

Knowing the material parameters V_0, σ_u, m, the probability of failure is an increasing function of the volume V and of course an increasing function of the damage equivalent stress σ^*.
As an application, let us calculate the mean value of the damage equivalent stress to fracture $\bar{\sigma}_R^*$ of a structure as a function of its volume V in the case of an uniform distribution of the initial damage

$$f(D_0) = \frac{1}{D_M} \quad 0 < D_0 < D_M,$$

$$P_{F_0} = \int_{D_c}^{D_M} \frac{\mathrm{d}D_0}{D_M} = 1 - \frac{D_c}{D_M},$$

with
$$D_c = 1 - \frac{\sigma_R^*}{\sigma_u}, \qquad P_{F0} = 1 - \frac{1}{D_M}\left(1 - \frac{\sigma_R^*}{\sigma_u}\right),$$

then
$$P_F = 1 - (1 - P_{F0})^{V/V_0} = 1 - \left[\frac{1}{D_M}\left(1 - \frac{\sigma_R^*}{\sigma_u}\right)\right]^{V/V_0}.$$

The mean value of the stress is given by

$$\bar{\sigma}_R^* = \int_0^\infty \sigma_R^* \frac{\partial P_F}{\partial \sigma_R^*} d\sigma_R^* .$$

Changing the variable σ_R^* for D_0 with

$$\sigma_R^* = \sigma_u (1 - D_0),$$

$$\bar{\sigma}_R^* = \int_0^{D_M} \sigma_u (1 - D_0) \frac{V}{V_0} \left(\frac{D_0}{D_M}\right)^{\frac{V - V_0}{V_0}} \frac{1}{\sigma_u D_M} \sigma_u \mathrm{d}D_0 ,$$

$$\bar{\sigma}_R^* = \sigma_u \frac{V}{V_0} \int_0^{D_M} (1 - D_0) \left(\frac{D_0}{D_M}\right)^{\frac{V - V_0}{V_0}} \frac{\mathrm{d}D_0}{D_M} ,$$

$$\bar{\sigma}_R^* = \sigma_u \left(1 - \frac{V D_M}{V + V_0}\right).$$

Let us consider the normalized mean failure stress ratio

$$\frac{\bar{\sigma}_R^*(V)}{\bar{\sigma}_R^*(V_0)} = \frac{1 - \dfrac{D_M}{1 + V_0/V}}{1 - \dfrac{D_M}{2}} .$$

This result is plotted in Figure 4.36 with the asymptotic value

$$\frac{V}{V_0} \to \infty \quad \Rightarrow \quad \frac{\bar{\sigma}_R^*(V)}{\bar{\sigma}_R^*(V_0)} = \frac{1 - D_M}{1 - \dfrac{D_M}{2}} .$$

The main difficulty in a volume effect analysis is the identification of the elementary representative volume V_0. Practically, V_0 may be taken as the volume of the uniform stressed part of the specimens used to identify the constitutive equations of the material under consideration, but it is difficult to avoid a certain degree of subjectivity!

A better identification would be to determine the probability density of defects $f(D_0)$ which allows for the calculation of P_{F_0} if $D_c(\sigma^*)$ is known. Then from tests on a real structure P_F may be derived and V_0 calculated from

$$P_F = 1 - (1 - P_{F_0})^{V/V_0} .$$

But this procedure needs two large sets of experiments!!

4.5.6 Effect of Stress Heterogeneity

When a structure is subjected to a nonuniform stress field, the probability of failure depends upon the probability of having large defects located in the stress concentration zone. If a stress field has a high degree of heterogeneity, the proba-

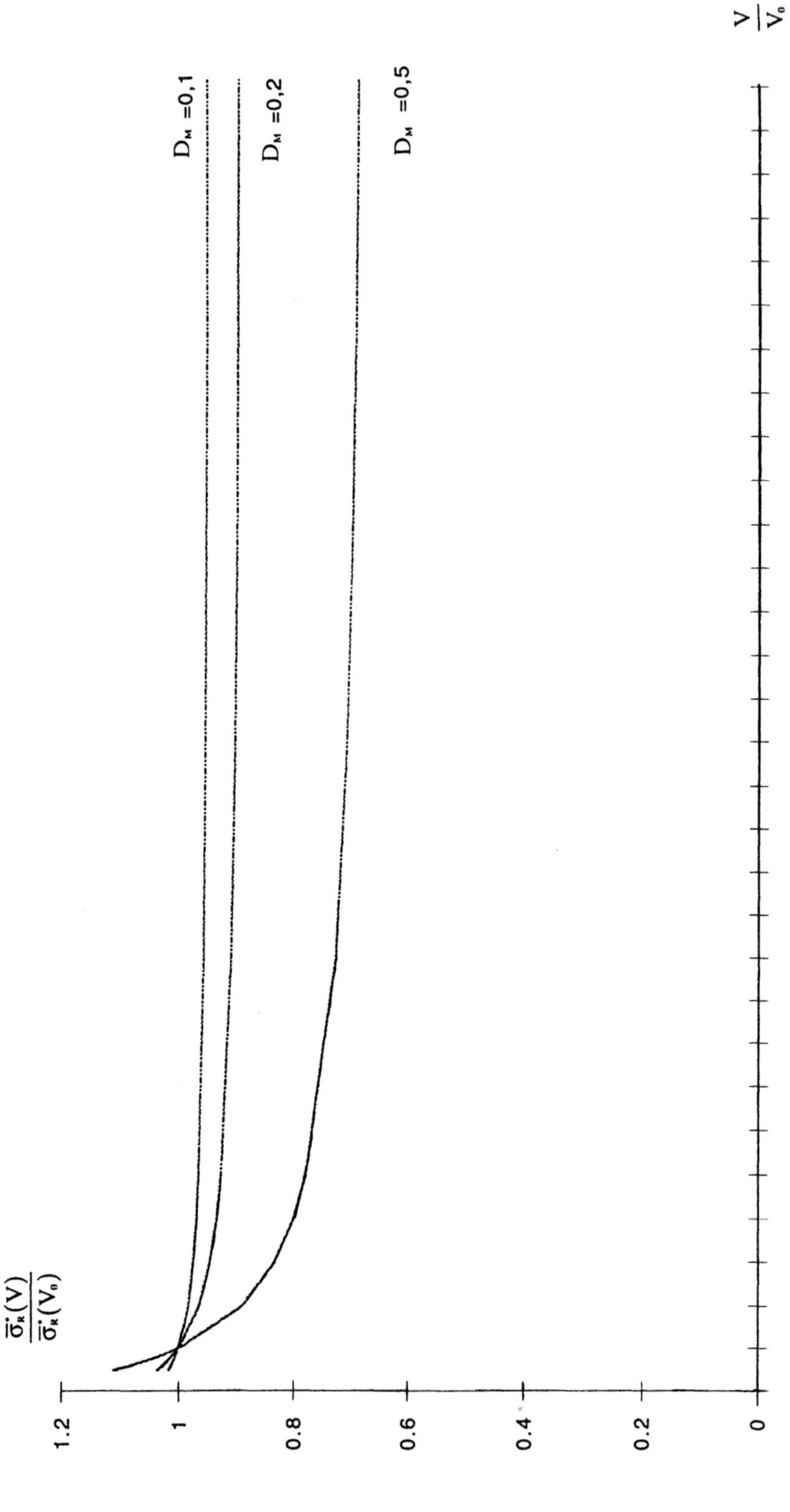

Fig. 4.36. Volume effect on the mean damaged equivalent stress at fracture

bility of failure for the same distribution of defects is larger than for a uniform stress field involving the same strain energy.

To model this effect, it is convenient to introduce the concept of effective volume V_{eff} defined by an equivalence in elastic strain energy as it governs the phenomenon of damage.

If w_e is the elastic strain energy density

w_e^{MAX} its maximum value at the stress concentration point of the structure,

$$\int_V w_{e(M)} \, \mathrm{d}V = V_{\text{eff}} \, w_e^{\text{MAX}},$$

or

$$\boxed{V_{\text{eff}} = \frac{\int_V w_{e(M)} \, \mathrm{d}V}{w_e^{\text{MAX}}}}.$$

Then, the probability of failure formulas are written with the effective volume calculated by an elastic calculation instead of the real volume of the structure,

$$P_F = 1 - (1 - P_{F_0})^{V_{\text{eff}}/V_0} \approx 1 - \exp\left(\frac{V_{\text{eff}}}{V_0} P_{F_0}\right),$$

or for the Weibull formula

$$P_F = 1 - \exp\left(-\frac{V_{\text{eff}}}{V_0}\left[\frac{\sigma^*_{\text{MAX}}}{\sigma_u}\right]^m\right).$$

4.5.7 Exercise on Bending Fatigue of a Beam

The considered beam is of rectangular cross-section and subjected to a pure periodical moment $M_{\text{MAX}} = M$, $M_{\min} = -M$ (Figure 4.37)

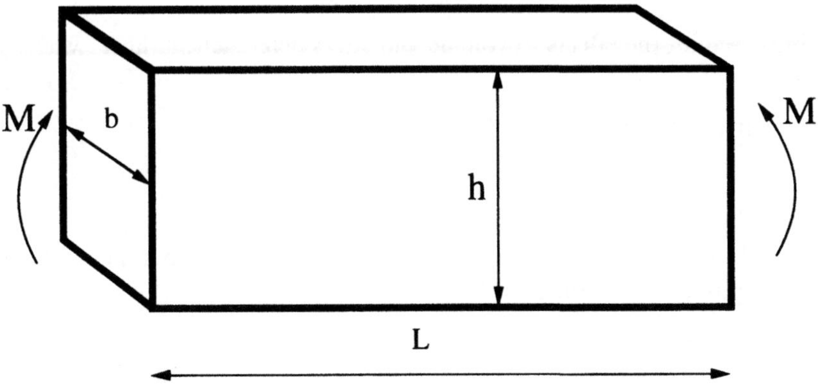

Fig. 4.37. Beam in bending fatigue

By a classical deterministic elastic analysis show that

— The one dimensional maximum stress is

$$\sigma_{\text{MAX}} = \frac{6M}{bh^2}.$$

— The strain amplitude is

$$\Delta\varepsilon = \frac{12M}{Ebh^2}.$$

— The number of cycles to failure from Section 4.5.3 is

$$N_R = \frac{ES}{\sigma_f^2 \Delta\varepsilon}(D_c - D_0).$$

By a statistical analysis considering a probability density of initial damage $f(D_0)$ and with

$$f_N(D)\mathrm{d}D = f(D_0)\mathrm{d}D_0,$$

show that:

— The probability of crack initiation is

$$P_{F_0} = \int_{(D_0)_c}^{D_M} f(D_0)\mathrm{d}D_0,$$

$(D_0)_c$ being the initial damage becoming critical after N_R cycles

$$(D_0)_c = D_c - \frac{N_R \sigma_f^2 \Delta\varepsilon}{ES}.$$

— For an uniform distribution

$$P_{F_0} = 1 - \frac{D_c}{D_M} + \frac{N_R \sigma_f^2 \Delta\varepsilon}{ESD_M}.$$

Taking into account the effects fo volume and stress heterogeneity linear with x_2 show that:

— The effective volume is

$$V_{\text{eff}} = \frac{Lbh}{3}.$$

Notice that the effective volume is reduced to Lbh/6 if pure unilateral conditions are considered that is no damage in compression.

— The volume ratio is

$$\frac{V_{\text{eff}}}{V_0} = \frac{1}{3},$$

if the constitutive equations of the material are identified by tests in tension on specimens of the same size as the beam.

— The probability of crack initiation of the beam in bending is

$$P_F = 1 - (1 - P_{F_0})^{V_{\text{eff}}/V_0} = 1 - \left(\frac{D_c}{D_M} - \frac{N_R \sigma_f^2 \Delta\varepsilon}{E S D_M} \right)^{1/3}.$$

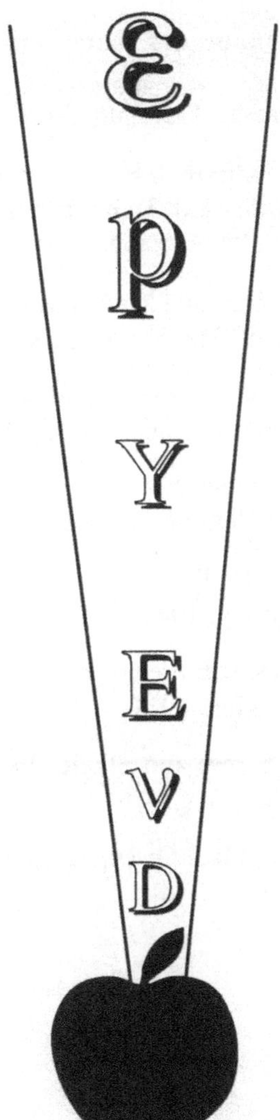

Bibliography

Atlas of Metal Damage. L. Engel, H. Klingele,
Wolfe Science Books, C.H. Verlag, Munich, 1981.

Mechanics of Solid Materials. J. Lemaitre, J.L. Chaboche,
Cambridge University Press, 1990
Translated from "**Mécanique des matériaux solides**," 2nd édition, 1988, Dunod 1985
Also translated in Chinese by T.C. Yu, 1991

Mechanics of Damage and Fatigue. S.R. Bodner and Z. Hashin,
Proceedings Symposium IUTAM, Haifa, Israel, 1985.
Pergamon Press, 1986.

Introduction to Continuum Damage Mechanics. L.M. Kachanov,
Martinus Nijhoff Dordrecht, The Netherlands, 1986.

Continuum Damage Mechanics – Theory and Applications. D. Krajcinovic,
J. Lemaitre, CISM Lectures, Springer Verlag, 1987.

Damage Mechanics. D. Krajcinovic,
in Mechanics of Materials Journal, 8, Elsevier, 1980.

Formulas for Stress and Strain. R.J. Roark,
McGraw Hill Book Company, 1965.

Stress Concentration Factors. R.E. Peterson,
Wiley Interscience Publication, 1974.

Handbook of Formulas for Stress and Strain. W. Griffel,
Frederick Ungar Publishing Co., 1976.

Finite elements in plasticity. D.R.J. Owen, E. Hinton,
Pineridge Press Ltd, 1980.

History of International Damage Mechanics Conferences

Damage Mechanics, Euromech Colloquium, Cachan, France, 1981.

Mechanics of Damage and Fracture, NSF Workshop, Atlanta, USA, 1982.

Mechanics of Damage and Fatigue, IUTAM Symposium, Haifa, Israel, 1985.

Continuum Damage Mechanics, CISM Lectures, Udine, Italy, 1986.

Large Deformations and Damage, Mini Symposium, IUTAM Congress, Grenoble, France 1988.

Strain Localization and Size Effects Due to Cracking and Damage, CNRS-NSF Workshop, Cachan, France, 1988.

Applications of Continuum Damage Mechanics, Euromech Colloquium, Crakow, Poland, 1989.

Damage Mechanics in Composite Materials Design, Santa Barbara UCSB Workshop, USA 1990.

Continuum Damage Mechanics, Sandia National Lab Workshop, Pleasanton, USA, 1990.

Damage Mechanics in Engineering Materials, ASME Winter Annual Meeting, Dallas (USA), 1990

Recent Advances in Damage Mechanics and Plasticity, ASME Summer Mechanics and Materials Conference, Tempe (USA), 1992

Computer aided Assessment and Control of Localized Damage, CAACOLD'92, Southampton (UK), 1992

Fracture and Damage of Concrete and Rock, FDCR 2, Vienna (Austria), 1992

Impact Damage in Composites, Euromech Colloquium, Oxford (UK), 1993

Computer Aided Assessment and Control of Localized Damage, CAACOLD'94, Udine (Italy), 1994

Structural Damage Assessment using Advanced Signal Processing Procedures, Int. Workshop, Pescara (Italy), 1995

Inelastic Deformation, Damage and Life Analysis, Int. Symposium. Hawaï (USA), 1995

Micromechanics of Plasticity and Damage of Multiphase Materials, IUTAM Symposium, Paris (France), 1995

Mechanical Behaviour of Damaged Solids, Mecamat Workshop, Fontainebleau (France), 1995

Author and Subject Index

A 542 stainless steel, 7
ABAQUS, 202
accumulated plastic strain, 52
acoustic emission, 35
AISI 1010 steel, 6, 28, 29, 33, 133
AISI 316 stainless steel, 32, 111, 139
Allix, 65
anisotropic damage, 59, 75, 80
associated variable, 40, 42

back stress, 14, 40, 50
Baptiste, 46
Bazant, 206
Beaujolais, 35
Belyschko, 206
Benallal, viii, ix, 7, 55, 165
Beranger, 214
Bernoulli, vii
Berthaud, 10, 26
Billardon, viii, ix, 28, 29, 86, 165, 203, 206
boundary value problem, 155
Bourgogne, 35
Bridgman's law, 31
brittle damage, 4, 68, 91, 112, 113, 197
Budiansky, 14, 42

Cailletaud, 32
carbon-carbon composite, 23
Cauchy stress tensor, 39
ceramics, 2, 3, 5, 112
Chaboche, viii, 43, 63, 206
characteristic length, 70
Chrzanowski, 90
Clausius-Duhem inequality, 41, 46, 99
cleavage, 4
composites, 2, 3, 11, 112
concrete, 2, 3, 24, 26, 112
consistancy condition, 53, 105, 155
constitutive equations, 55, 107, 155
convergence, 166
copper, 22
Cordebois, 121
Coulomb's coefficient, 77
coupled constitutive equations, 50
coupling, 14
crack closure, 19
crack closure parameter, 80, 111
crack initiation, 102, 115, 169, 172
creep damage, 4, 33, 123
creep fatigue interaction, 132
creep rupture, 10, 91
critical damage, 98, 103
critical point(s), 169
critical time, 162

DAMAGE, 90, 179
damage accumulation, 141
damage equivalent stress, 44, 85
damage growth, 70, 73, 177
damage localization, 86
damage measurement, 35
damage orthotropic, 62
damage threshold, 16, 98, 108
damage variable, 11
damaged zones, 205
debonding of interface, 208, 212
decohesion, 16
deep drawing, 119
density, 29
deviatoric stress, 65
discontinuity, 86
dissipation, 47
Doghri, ix, 86, 165, 177, 179, 206
double scalar variable, 64
ductile damage, 4, 27, 72, 117, 196
ductile fracture, 11
Dufailly, 6, 8, 15, 22, 29, 111

effective damage area, 17
effective stress, 12, 19, 42, 62, 84
elastic predictor, 166
elasticity modulus, 14, 21
electrical resistance, 30
energy strength of damage, 98
Engel, 6, 7, 8, 9, 10, 20, 21
equilibrium equations, 155
Eshelby, 76
Euler's angles, 75
Evans, viii
extrusion, 119

fatigue crack, 11
fatigue limit, 109
Fenchel transform, 78
fiber bundle system, 90
fiber composite, 65
finite element method, 165
first principle of thermodynamics, 101
forging, 119
fourth-order damage tensor, 63
fracture limits, 119
François, viii, 10
free energy, 46
frightened apple, 208
fully coupled analysis, 192

Gauss point, 166
generalized standard materials, 47
Germain, viii
Geymonat, 86
Gibbs specific enthalpy, 83
Goodman's diagram, 139
Gunawardena, ix
Gurson model, 72, 74

Hayhurst, 60, 206
Heaviside, xv, xvi
Helmholtz free energy, 41
Herath, ix
high cycle fatigue, 7, 27, 135, 197
high-alloy steel, 19
Hild, 5, 128, 209
homogenization, 69
Hult, viii, 90
Hutchinson, viii, 208
hydrostativ stress, 65

identification, 57, 109, 110
implicit integration, 166
implicit scheme, 178
IN 100 superalloy, 32
inclusions, 4
incremental linearization, 165
initial conditions, 155
initial damage, 179, 202
initial plastic strain, 179
initial strain hardening, 202
input data, 180
instability, 86
interface, 66, 208
intergranular decohesions, 5
intergranular microcracking, 7
internal variables, 40
isochronous creep damage curves, 126
isotropic damage, 39
isotropic hardening, 14, 40, 42, 51

jump-in-cycles procedure, 166, 179
June Finney, ix

Kachanov, v, vii, 11
Kachanov's model, 124
kinematic hardening, 14, 40, 43, 50
kinetic damage law, 169
kinetic energy, 101
kinetic law of damage, 47, 48
Klingele, 6, 7, 8, 9, 10, 20, 21
Krajcinovic, viii, 2, 75, 76
Kronecker, xv

Ladeveze, 64, 83
Lamé's constants, 79
Leckie, viii, ix, 60, 79
Lee Hao, vii
Legendre-Fenchel, 85
Lemaitre, v, 11, 13, 39, 41, 95, 176
Lienard, 20, 21, 168
Lin-Taylor hypothesis, 81, 114, 136, 177
linear accumulation, 142
linear creep fatigue interaction, 135
Lippmann, v, vii, viii, ix
listing of DAMAGE, 90, 185
loading function, 48, 49
loading-unloading condition, 49
localization, 86, 176
locally coupled analysis, 176
low cycle fatigue, 5, 27, 127, 197
lower bound, 171
Luder's bands, 89

Macauley, xv
macrocracks, 205
macroscale, 4
Manderscheid, ix
Manson-Coffin law, 131
MAR M 509, 36
Marquis, viii, ix, 39
master curves, 120
material parameters, 57, 98, 104, 109, 152
Maxwell's compatibility equation, 87
Mazars, 24
McClintock, 74
McMeeking, viii
mean stress, 138
measurements, 19
mechanism of damage, 10
mesoscale, 4, 116
metal forming, 119
metals, 1, 3, 117
micro crack closure effect, 107
micro-meso element, 68, 72
microcavities, 72
microcrack, 2, 19, 68

microcrack closure effect, 80
micrograph pictures, 19
micrographic observations, 10
microhardness, 25
micromechanics, 68
microscale, 4, 116
Miner, vii
Moussy, 30
multiaxial fatigue, 200
multihole, 159
multiplier, 49
Mura, 76
Murakami, vii, viii, 60

Najar, viii
Needleman, 206
Neuber's method, 163
Newton's procedure, 166, 178
nonlinear interaction, 132
nonlinear kinematic hardening, 51
normality condition, 47
Norton, xvii
Norton's law, 33, 49, 124
notches, 156
nucleation of microcracks, 5
numerical simulation, 196

O'Connell, 14, 42
observable variables, 39
Ohm's law, 30
Onat, 79
Ortiz, 206
output data, 183

Palmgreen, vii
Palmgreen-Miner's rule, 141
Paris, xvi, xviii
Paris law, 70
perfectly plastic materials, 105
Pijaudier-Cabot, 206
Pineau, viii
plane strain, 66
plastic corrector, 166
plasticity, 3, 14, 48, 53
plasticity criterion, 105
Poisson, xviii
Poisson's ratio, 18, 42
Policella, 34
polymers, 2, 3, 117
Poss, 23
postprocessing, 177
potential drop, 30
potential of dissipation, 46, 95
proportional loading, 66, 117

quasi-brittle damage, 113, 135

Rabotnov, vii, 12
Rayleigh's surface waves, 25
representative volume element, 3, 11
residual stresses, 203
Rice, 208
Rice and Tracey, 74
Robinson, vii
rolling, 119
Rosengreen, 208
rupture criterion, 16, 100, 108
rupture domain, 103

Saanouni, 206
Sabine Lemaitre, ix
safe domain, 103
Sampaio, vii
scale, 3
second-order damage tensor, 60
Sermage, 151
Sherman, ix
slips, 2, 3
state kinetic coupling theory, 42, 96, 210
state potential, 39
strain energy density release rate, 43, 107
strain energy release rate, 18, 69
strain equivalence, 13
strain hardening, 27
strain measurements, 21
strain-displacement equations, 155
strain-driven algorithm, 178
stress amplitude drop, 31, 36
stress concentration, 17, 156, 167
stress concentration factor, 156
Stress corrosion cracking, 11
stress intensity factor, 18
summation convention of Einstein, 40
SYSTUS, 177, 202

Taira rule, 135
tangent modulus, 166
tangent stiffness tensor, 86
Taylor, 78
tensile crack, 11
thermal cracks, 11
thermodynamics, 39
threshold, 98
transcristalline cleavage, 10
transcristalline fatigue crack, 10
transgranular microcracking, 8
triaxiality, 116
triaxiality function, 45, 67
triaxiality ratio, 44, 67, 113
Tvergaard, 74
two-scale volume element, 114

ultimate stress, 16, 109
ultraconic waves, 23

uncoupled analysis, 169
unified damage laws, 95
unilateral conditions, 80

viscoplasticity, 49, 55, 108
von Mises, xix
von Mises equivalent stress, 40, 44

Weibull, v, 216
wine selection, 35
Woehler curve, 138
wood, 2-3

Young, xvi
Young's modulus, 19, 42

Zaoui, viii

GPSR Compliance

The European Union's (EU) General Product Safety Regulation (GPSR) is a set of rules that requires consumer products to be safe and our obligations to ensure this.

If you have any concerns about our products, you can contact us on ProductSafety@springernature.com

In case Publisher is established outside the EU, the EU authorized representative is:

Springer Nature Customer Service Center GmbH
Europaplatz 3
69115 Heidelberg, Germany

Batch number: 10704713

Printed by Printforce, the Netherlands